AF601247

Advances in Bioclimatology 3

Volumes in the series

Vol. 1: R.L. Desjardins, R.M. Gifford, T. Nilson, E.A.N. Greenwood
(1992)

Vol. 2: J.D. Kalma, G.P. Laughlin, J.M. Caprio, P.J.C. Hamer
The Bioclimatology of Frost. Its Occurrence, Impact and Protection
(1992)

Vol. 3: Y. Cohen, J.M. Elwood, M.G. Holmes, V.A. Kanevski, S.M. Kochubey, J. Ross, T. Shadchina, E. Simensen, F.I. Woodward
(1994)

Advances in Bioclimatology_3

With Contributions by

Y. Cohen J.M. Elwood M.G. Holmes V.A. Kanevski
S.M. Kochubey J. Ross T. Shadchina E. Simensen
F.I. Woodward

With 60 Figures

Springer-Verlag Berlin Heidelberg GmbH

Edited by
Dr. GERALD STANHILL
Agricultural Research Organization
The Volcani Center
Institute of Soils and Water
P.O. Box 6
Bet Dagan, 50250
Israel

ISBN 978-3-540-56381-5 ISBN 978-3-642-57966-0 (eBook)
DOI 10.1007/978-3-642-57966-0

Library of Congress Cataloging-in-Publication Data.

Originally published by Springer-Verlag Berlin Heidelberg New York in 1994

Typesetting: Macmillan India Ltd., Bangalore-25
31/3130/SPS - 5 4 3 2 1 0 - Printed on acid-free paper

Preface to the Series

Advances in Bioclimatology – the study of the relations between the physical environment and the form and function of living organisms – have been spectacular during the last third of this century.

Before this period, the subject, having slowly emerged from its classical origins as a branch of natural history, had reached the stage of a collection of largely empirical, statistical relationships between standardized but often inappropriate climatological and biological measurements.

Since then, research into the basic physical and physiological mechanisms involved has used the latest techniques of measurement and analysis to develop various bioclimatic relations which have contributed much to improving crop and animal production and optimizing the human environment.

Recently, some of these relationships have been incorporated into larger models of climate-ecosystem interactions. Such models are being used to assess the often unintended effects of human activity on various elements of the biosphere.

However, the advances described have been very unevenly spread through the vast field of interest encompassed by bioclimatology; the fields of plant, animal and human climatology have largely advanced in independent fashions and even within each biological province different techniques of analysis and measurement have developed for different time and space scales of organization.

One of the major aims of this new review series is to overcome this separate development by providing a common forum for those wishing to obtain an authoritative review of the latest developments in bioclimatology.

The emphasis will be on advances which are soundly based on biological and physical principles rather than those describing empirical relationships. Reviews will also deal with the latest techniques of measurement and analysis which are of relevance to bioclimatology and to those describing broader ecological studies in which bioclimatology provides a major element.

Although most of the reviews to be published will be commissioned, the editors would welcome suggestions from individuals interested in contributing a review of the type described, as well as for ideas on major topics of wide interest around which a number of individual reviews could be centered.

Bet Dagan, Israel

G. STANHILL
Editor

List of Editors

Managing Editor

Those interested in contributing a review to this series are invited to contact Dr. Gerald Stanhill or one of the Associate Editors. Proposals should outline briefly the review's aims and scope.

Contents

1 Human Melanoma and Ultraviolet Radiation

J.M. Elwood

1 Introduction

Melanoma is the most important type of skin cancer. For the moderate incidence area of England and Wales (population 60 million), there were approximately 1000 deaths from melanoma in 1985, over twice as many as from all other skin cancers combined. For a population of 1 million people in the United Kingdom, there are about 45 new cases of melanoma, compared to some 450 registered non-melanoma skin cancers. Incidence data for non-melanoma skin cancer may suffer from substantial under-reporting. However, whereas non-melanoma skin cancer affects elderly people and has a high cure rate, melanoma affects a relatively young population and is a serious tumour with a substantial mortality; almost 40% of patients die from the disease within 5 years. The deaths per million are therefore about 20 per year for melanoma, compared to 10 for other skin cancers (Office of Population Census and Surveys 1980, 1987, 1988).

1.1 Factors Influencing the Risk of Melanoma

An individual's risk of melanoma depends on two sets of factors: characteristics of the individual such as his or her racial origin, pigmentation, reaction to sunlight, and other skin features such as naevi; and characteristics of the environment in which he or she lives. The well-established external causal factor for melanoma is sun exposure, with there being considerable evidence for a complementary effect of artificial ultraviolet radiation. In this review, the emphasis is on the role of ultraviolet radiation as a cause of melanoma, and some information previously published is brought together and updated (Elwood 1984, 1989a, b; Elwood et al. 1989). Other reviews are available which collate the information on the importance of intrinsic factors in determining an individual's risk (Armstrong and English 1990).

2 Melanoma and Solar Ultraviolet Radiation

2.1 Evidence for a Causal Role of Sun Exposure in Human Cutaneous Melanoma

The evidence for a causal effect of sun exposure on cutaneous melanoma comes from five sources; general characteristics of the tumour, experimental data, the analogy with non-melanoma skin cancer, the association of melanoma with xeroderma pigmentosum and similar states, and, the most important, human epidemiological data.

The first line of reasoning is simplistic. Just as cancer of the gastrointestinal tract is likely to be related to dietary intake, and cancer of the respiratory tract to contaminants carried in inspired air, it is logical to look for causes of malignancy of melanocytes to the agent they are designed to respond to, that is ultraviolet radiation.

Animal data directly relevant to melanoma have until recently been sparse, with no clearly analogous animal model on which extensive studies have been done. Kripke in 1979 suggested four mechanisms by which ultraviolet radiation could produce melanoma. First, an initiation action, with no other influences, for which there is no well-established animal model. Second, an initiation of effect requiring additional promotion before expression, documented at that time by only one instance in which a mouse given ultraviolet radiation followed by croton oil treatment produced a melanoma. Third, ultraviolet radiation acting as a promoter on cells already initiated. There is much more evidence for this, with melanoma in guinea pigs and hamsters being produced by chemical initiators followed by a promoter action of ultraviolet radiation (Beattie et al. 1988). Fourth, an indirect promoter action via immunological suppression, as demonstrated by Kripke's elegant experiments on mouse melanoma (Fisher and Kripke 1977, 1978). This mechanism, unlike the others, implies that the action of ultraviolet radiation is not confined to initiated melanocytes exposed directly to the radiation, but includes effects on melanocytes in other body sites not directly exposed.

Thus, until recently, there has been no reliable animal model in which UVR (UV radiation) has been used as a sole agent or an initiator in the production of melanoma. However, mitogenic effects of UVR on normal human melanocytes in culture have now been shown (Libow et al. 1988), and recently Ley et al. (1989) have reported the induction of melanoma by UVR alone in the South American opossum, *Monodelphis domestica*; and Setlow et al. (1989) have produced melanomas in a platyfish-swordtail hybrid by UVR; these new models may be useful. Results most relevant to the human situation will arise if a model suitable for the study of differing dosages, wavelengths, dose rates and fractionations can be developed.

Assistance with the understanding of the causes of melanoma is given by knowledge about non-melanoma skin cancer, particularly squamous cell carcinoma, which is one of the most clearly understood tumours in regard to its causation. Animal data are extremely extensive, and the production of squamous cell tumours by ultraviolet radiation in mice and other animals has been one of the most heavily researched areas of animal carcinogenesis (Urbach et al. 1974; National Research Council 1979, 1982). A single or short exposure to ultraviolet radiation will not in itself produce tumors; the minimum recorded latent period from first exposure in mouse skin experiments is 74 days following a considerable dose over a protracted period; however, papillomas have been seen to arise at the edge of an area of mouse skin severely traumatised by short heavy doses of ultraviolet radiation. The yield from ultraviolet exposure is increased by the application of promoters such as croton oil, and can also be increased by heat, humidity, and wind. Tumours have mainly been produced by ultraviolet B radiation in the 280–320 nm range, but the addition of ultraviolet A increases the yield. For a given total dose of ultraviolet radiation (measured by intensity multiplied by time of exposure), tumour yield is greater if the dose is given in many repeated portions, for example daily fractionation, rather than by more intensive exposures less frequently, although there are some variations in these results between different species. The animal data are consistent with the concept that the probability, or risk, of producing squamous cell tumours is a simple function of the total dose of ultraviolet radiation received (Blum 1959).

The descriptive epidemiological evidence in relation to non-melanoma skin cancers is also relatively simple, and appears consistent with a model of risk being due to total cumulated ultraviolet dose on that particular skin site, over the lifetime. Thus non-melanoma skin cancers are virtually restricted to extensively exposed skin areas, and show a linear log incidence–log age curve with a slope similar to that of other epithelial tumours. Moreover, they show increased incidence rates at lower latitudes within relatively homogeneous populations, a strong association with outdoor employment which occurs indirectly as a male excess in most countries, and an association with lower rather than higher socio-economic status (National Research Council 1982). Non-melanoma skin cancers are more common in individuals with relatively light skin and hair colour, and with ethnic backgrounds which are associated with such factors, such as Irish or Scottish ancestry. Although non-melanoma skin cancers can also be produced by a variety of chemical carcinogens (e.g. DMBA), it is likely that the great majority of these tumours in human populations are the result of accumulated high dosage of ultraviolet radiation over a lifetime. As a result, where relatively pale-skinned populations have intense sun exposure, as in the British origin population in Australia, non-melanoma skin cancer is very common (Marks et al. 1983). However, a recent case–control study suggests that the dose relationship is not simple, and intermittent exposure may be involved (Kricker et al. 1991a, b). Xeroderma pigmentosum is a rare autosomal recessive condition showing severe skin damage, including great susceptibility to sunburn, premature aging, and a great increase (more than 1000 times) in the risk of non-melanoma skin cancer (Kraemer et al. 1987; Arlett and Cole 1989). The mechanism linking UV to skin cancer has been studied primarily in relationship to xeroderma pigmentosum and non-melanoma skin cancer, with little work directly related to melanoma. At the DNA level, UVR is absorbed by purines and pyrimidines, allowing dimers to form which link the DNA strands, blocking replication and transcription, and being potentially mutagenic. Such damage is usually reversed by an incision and rejoining process: specific endonucleases can alleviate the defect of repair shown by xeroderma pigmentosum cells (Hanawalt et al. 1979). An action spectrum for pyrimidine dimer formation in human skin shows a peak near 300 nm, the rapid decrease at shorter wavelengths being due to absorption of UV by the upper layers of the skin. An estimate of a 2.5-fold increase in dimer formation for a 50% depletion of ozone has been made (Freeman et al. 1989). Nine subgroups of xeroderma pigmentosum have been identified (Lancet 1989). However, another autosomal recessive condition, Cockayne's syndrome, shows susceptibility to DNA damage but no increased cancer risk (Lehmann 1987); it is suggested that skin cancer requires both the defect in DNA repair and defective immune surveillance (Lancet 1989; Bridges 1981).

2.2 Epidemiology of Melanoma

The general epidemiological features of melanoma became apparent from descriptive studies by the mid-1970s. The distinction of melanoma into subtypes is important. In white populations, some 80–90% of cutaneous melanoma is of the superficial spreading and nodular types, and so it is the evidence on these major subtypes which will be emphasized; the main features which will be further discussed are listed in Table 1, showing the contrast to the descriptive data on non-melanoma skin cancer. Lentigo maligna melanoma shares most of the epidemiological features of non-melanoma skin cancer, being concentrated on exposed sites and showing a linear log incidence–log age relationship, and therefore the risk is likely to be a direct function of cumulated direct ultraviolet dosage to that body site (Newell et al. 1988). Acro-lentiginous melanoma is also likely to have a very different aetiology, being a relatively common tumour in Oriental populations, and showing a site distribution which appears independent of sun exposure (Elwood 1989a).

2.3 Geography

There is a clear latitudinal gradient of melanoma within relatively homogeneous populations, such as those of Australia, the United States and Canada, England and Wales, and Scandinavia (Elwood et al. 1974; Lee 1982; Jensen and Osterlind 1988; Fig. 1). The proportional increase with latitude is similar to that of non-melanoma skin cancer, and the change in rate with latitude is more rapid for females than males (National Research Council 1982). Migrant studies show that subjects who move from less sunny places to countries such as Australia, Israel, and California show lower rates than the native population, with one exception which is that Caucasian migrants to Hawaii show higher rates of melanoma than the locally born population (Elwood 1984).

2.4 Socio-Economic Factors

The other descriptive features in melanoma stand in contrast to the epidemiology of non-melanoma skin cancer, showing that the aetiology must be different. Melanoma

Table 1. Comparison of non-melanoma skin cancer and melanoma[a]

Feature	NMSC	Melanoma[a]
Mainly in white populations	Yes	Yes
Increased with light pigmentation	Yes	Yes
Increased at lower latitudes	Yes	Yes
Strong concentration on exposed sites	Yes	No
Male excess	Yes	No
Increased in outdoor workers	Yes	No
Linear log-incidence log-age relationship	Yes	No
Rates increasing	?	Yes
Risk proportional to cumulated dose	?	No

[a] Cutaneous melanoma, superficial spreading and nodular types.

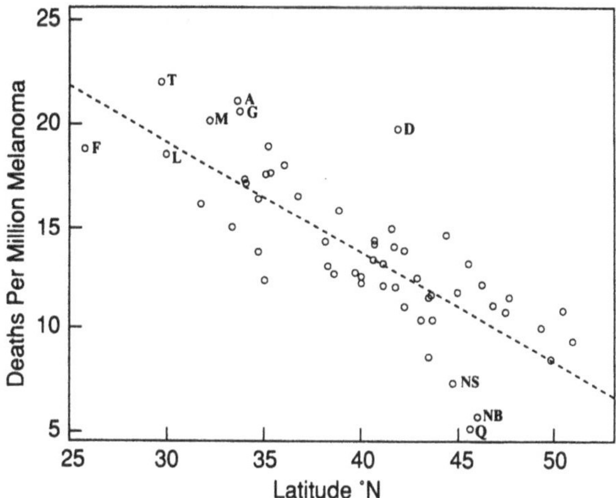

Fig. 1. Relationship of age-standardised male mortality rates from melanoma for each US state and Canadian province, 1950–1967, plotted against the latitude of the largest city. *F* Florida; *A* Alabama; *NB* New Brunswick; *D* Delaware; *Q* Quebec; *T* Texas; *G* Georgia; *L* Louisiana; *M* Mississippi; *NS* Nova Scotia. Newfoundland not shown (mortality 1.93, latitude 47.6°). (Elwood et al. 1974)

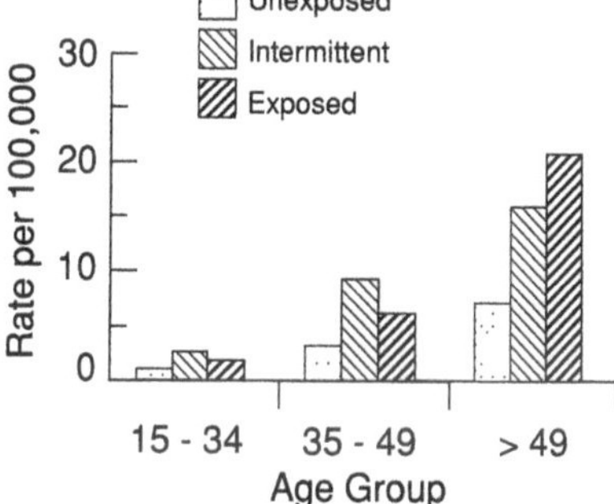

Fig. 2. Incidence rates of melanoma expressed as cases per 1 000 000 population per year, per unit area of skin, for skin sites divided by usual sun exposure from a continuous series of 281 new cases of cutaneous melanoma of the superficial spreading or nodular type, seen in Vancouver, Canada from 1976 to 1979. (Elwood and Gallagher 1983)

is much more common in the richer rather than the poorer socio-economic groups, and in comparing occupational groups of similar socio-economic status, melanoma rates are lower in groups with outdoor occupations than in those with indoor occupations (Lee and Strickland 1980; Elwood 1984).

2.5 Site Distribution

The body site distribution of melanoma is different from non-melanoma skin cancer. Analysis of the incidence rate of melanoma per unit area of skin can compare body sites which are usually exposed such as the face, usually unexposed such as the lower back and buttocks, and sites which are intermittently exposed such as the legs and the back in men. Incidence rates per unit area are clearly much lower on the unexposed sites (Fig. 2). However, they are similar on exposed and on intermittently exposed sites; in Canadian data in younger patients the incidence rates were maximum on intermittently exposed sites, whereas in older patients they were higher on continuously exposed sites (Elwood and Gallagher 1983). The total cumulative dose of UVR at an intermittently exposed site can never exceed that of the totally exposed site. This similarity of the incidence rates per unit area must therefore mean that intermittent exposure for the same total dose has a greater effect on melanoma risk than continuous exposure. An alternative explanation is that the risk–dose relationship reaches a maximum fairly early so that an intermittently exposed site has sufficient dose to reach maximum risk, and the extra dosage from continuous exposure adds nothing; although this is ruled out if the rates are truly higher on intermittently than on continuously exposed sites. It would be good to have such analyses which take account of skin thickness, pigmentation, and variations in melanocyte density.

The site distribution of melanoma has not been constant but has changed considerably over time (Elwood and Lee 1975). Study of the increase in melanoma shows that this has been most marked for the trunk in the male, and the lower limb in the female, so that the characteristic distribution of melanoma with high rates on these two sites is a relatively recent phenomenon. Whereas the increasing time trend shows this clear site specificity, the geographical trend is not clearly site-specific; places nearer the equator have higher rates of melanoma at all sites. In this regard, it may be relevant that UVR induces an increase in melanocytes in human skin in shaded areas as well as exposed areas (Stierner et al. 1989).

2.6 Time Trend in Frequency

The other major feature of melanoma incidence of course is the rapidly increasing incidence rate seen in all populations over the last few decades (Muir and Nectoux 1982). There is no evidence as yet that we have seen maximum rates in any population; in fact, in a recent analysis, the most rapidly increasing rate was seen in the Caucasian population of Hawaii, which also had the highest recorded rate (Armstrong 1988). One of the challenges of aetiological studies is the question of whether the causal factors identified are an adequate explanation of this remarkable rise in incidence.

2.7 The Intermittent Exposure Hypothesis

The data available by the mid-1970s gave rise to the intermittent exposure hypothesis, based on the descriptive features (Elwood and Hislop 1982; Holman et al. 1983). This hypothesis was tested in a number of epidemiological studies in which a series of melanoma patients were compared to comparison subjects and standardised questions used to elicit past histories of behaviours and environmental characteristics which could be interpreted in terms of sun exposure. There have been more than 20 of these studies published to date, and there are excellent reviews available of all these studies (Armstrong 1988; Armstrong and English 1990; Elwood 1992; International Agency for Research on Cancer 1992).

2.8 The Case–Control Study

The case–control study design is the most powerful method of assessing causal factors which may require a long latent period of operation, and which are related to a disease which is still in population terms, relatively unusual. For example a case–control study for which information is gathered on 200 melanoma patients and 200 controls contains from a statistical point of view about the same amount of information as a prospective study carried out over a period of 20 years, involving the follow-up of some 100 000 people. This latter prospective study may still miss the relevant items if the latent period between critical exposure and disease occurrence is more than 20 years, and is obviously an almost impossible logistic undertaking. The critical issues in case–control study design relate to the three major characteristics of the design. First, the study involved selecting a representative series of melanoma patients. Second, these are compared to a group of subjects whose exposure to the putative causal factors such as sun exposure can be taken as representative of the population in which the cases have arisen. Third, the methods of collecting information must provide comparable and unbiased data for the cases and for the controls. The four studies on which the main evidence for solar exposure in melanoma depends are those performed in western Canada (Elwood et al. 1984, 1985a, b, 1987; Gallagher et al. 1986b, 1987), in Western Australia (Holman and Armstrong 1984a, b; Holman et al. 1986a, b, English and Armstrong 1988), Queensland (Green 1984; Green et al. 1985a, b, 1986), and in eastern Denmark (Osterlind et al. 1988a, b). They all share certain characteristics of high quality study design. All diagnosed patients with cutaneous melanoma in a defined population over a defined time period were identified and approached to be included in the study. The comparison groups of subjects were obtained by a random sample technique applied to the populations from which these cases had arisen, and therefore represent the normal healthy people of the same age and sex in that population. The methods used to obtain data rely on extensively tested, standardised questionnaires, administered by trained interviewers who (except in Queensland) were separate from the principal investigators and hypothesis-makers related to the studies. The relationship between a subject in such a study and an interviewer is a non-threatening relationship often quite different to the relationship between that subject and the doctor treating them. The essential objective of such methods is to obtain information which is unbiased, i.e. similar in accuracy in both the cases and controls.

2.9 Interpretation of Case–Control Studies

The case–control study design has been an important part of cancer aetiology studies since the early 1950s, when the studies relating lung cancer to smoking were published. The results of the British case–control studies of lung cancer, relying on relatively small numbers of patients and comparison subjects and published in the early 1950s, compared well with the results from the prospective follow-up studies in the same country which reported 10 and even 20 years later, but involved the study of a much greater number of subjects (Elwood 1988). This does not mean that all case–control study results are as consistent, and there are of course many examples of case–control studies producing results which have not been confirmed subsequently.

The basic result of a case–control study is a measurement of the association between an ‘exposure’ such as sun exposure, and an outcome such as melanoma. This is expressed as the relative risk; a relative risk of 3 means that the risk of melanoma in the 'exposed' group is three times that of the 'reference' group used in the comparison.

Where an association such as a link between solar exposure and melanoma is produced by a case–control study we must consider four general possible explanations of that association. First, of course, the association may represent causality; we arrive at this conclusion only after excluding the other possibilities. Second, the association may be due to chance variation; we can assess the probability of that by standard statistical techniques.

The other two possibilities are more difficult to deal with. One is that the association is due to bias in the data collected. This is a major consideration in these melanoma studies. The studies use complex questionnaire techniques to obtain information on the individual's sun exposure over many years. The questionnaires are detailed enough to distinguish different types of exposure, for example exposure from occupation and exposure from recreational and vacational activities. There is clearly a large amount of *inaccuracy* in such data. If this inaccuracy applies equally to all subjects in the study, that is, is the same for the melanoma patients as for the controls, its effect on the study results will be to make it more difficult to observe an association where a true association exists. The random error will dilute any true association, and make the results of the study overly conservative. Thus, if an inaccurate measurement produces an association with a relative risk of 3, for example, the true association must be much stronger than that. What is of much more concern is whether there is bias, that is, have the melanoma patients responded differently to questions on sun exposure than the comparison subjects.

There are a number of checks we can use to assess this possibility. The primary method of avoidance of bias is to use questionnaires which are standardised so that the same question is asked in the same way and in the same order to all the subjects, the questions are non-threatening, and the interviewers (as far as possible) do not know the case or control status of the interviewee. Such methods will deal adequately with potential bias produced by the interviewer. However, they will not approach the major issue, which is that the melanoma patients know that they have melanoma (or at least some serious disease), and the controls will know that they are healthy; this is the major source of potential bias. The likelihood of bias will depend on the individual's level of knowledge and suspicions about the items questioned. Studies of melanoma and sun exposure are likely to be more difficult now, because there has been very extensive publicity in regard to the association. In the mid-1970s when we developed the western Canada melanoma study, the level of public awareness and knowledge of the problem was very different. There had been none of the publicity which has since occurred (as a result of these studies), so that the idea of melanoma as a common, serious disease was not recognised. Moreover, there had been virtually no publicity in Canada relating sun exposure to risks of skin cancer. There was a lower public awareness of the whole problem, and also a different professional attitude. Our experience was that most dermatologists we talked to in the early stages of development of that study dismissed the possibility of a relationship between sun exposure and melanoma, primarily on the grounds of the site distribution of melanoma. Amongst our colleagues in dermatology and clinical oncology, there were more who thought the hypothesis linking sun exposure and melanoma was so unlikely that it was not worth exploring, than there were who already accepted that there was a relationship. Patients with melanoma at that time were not given any advice about changes in sun behaviour. On the other hand, some of the other causes of cancer, particularly smoking, but also alcohol, dairy fats, and artificial sweeteners has had much publicity. Most of the lay public do not appreciate the, to them, subtle differences between cancer of different sites. This fact can be useful in assessing these studies. If there were extensive bias in the responses to the questionnaire, we would expect that in the absence of any true association, say with smoking, there would be more reporting of smoking in the cases than in the controls. But the results in the western Canada melanoma study show that there is no difference in many of the factors commonly associated in the public mind with cancer such as smoking, alcohol, and aspects of diet (Table 2). We would also expect that patients who had had a skin cancer would be particularly sensitive to questions relating to previous skin problems, but again there was no difference between cases and controls in the frequency of reporting frequent skin conditions, and frequent general medical conditions (Gallagher et al. 1986b). Such results make us more confident that an observed difference in reported sun exposure histories, which were less likely to be associated in the public mind with the disease, are unlikely to be due to bias in reporting.

Another important issue is whether the cases and comparison subjects involved in the study are truly representative. In the four major studies the cases are all diagnosed melanoma patients in a particular population in a known time period,

Table 2. Assessment of possible observation bias in case–control studies: results of questions which are likely to be open to recall bias (Western Canada Melanoma Study)

Exposures often regarded as cancer-related; percentage reporting use of:	Cases	Controls
Tobacco	64	63
Alcohol	82	80
Coffee	90	92
Artificial sweeteners	20	17
Beef > 3/week	75	76
Milk > 3/week	75	74
Percentage reporting history of:		
Other skin conditions		
Acne vulgaris	24	23
Psoriasis	6	5
Other medical conditions		
Rubella	68	61
Mumps	74	73
Diabetes	4	5
Hypertension	16	19
Arthritis	13	17

Expanded data based on Gallagher et al. (1986b)

and the comparison subjects are randomly selected from the source population. However, not all cases and not all controls finally end up in the study, as cases are excluded because they have died, because their physicians want them excluded from the study for various reasons, or because they do not wish to participate; and comparison subjects are excluded because they cannot be reached or because they do not wish to participate. The final participation rates of cases and controls in the various studies vary from around 60 to 100%. These figures are still of course much higher than the participation rate in, for example, a clinical trial, where the number of patients entered into the trial is likely to be an extremely small proportion of the total eligible patients in a given population. In the western Canada study we could look at various measures to assess if these losses would affect the study results. We can compare, for example, some characteristics of the cases of melanoma who were finally interviewed compared with all cases who could have been interviewed; these show that the characteristics are very similar: (Table 3). The major differences were in age and sex composition, which is not a problem as the age and sex of the comparison subjects is made to be the same as the cases. We had more concerns about our control population where the response rate was lower because of various administrative issues and confidentiality issues, but for a few factors we have information from census data on the same population; comparisons show that the interviewed controls are very similar to the total population from which they are drawn (Table 4). Of course we have no information on the questions of prime importance such as sun exposure in those we cannot involve in the study.

Table 3. Possible selection bias in case–control studies: effect of non-response on case series (comparison of interviewed and non-interviewed cases, percentage distribution; Western Canada Melanoma Study)

		Interviewed n = 651	All subjects n = 766
Age	< 34	28	27
	> 35–59	43	44
	60 +	28	30
Sex	Male	40	42
	Female	60	58
Tumour type[a]	SSM	64	63
	NM	20	20
	UM	8	8
	LMM	9	9
Clark level	1, 2	39	38
	3	24	23
	4, 5	29	28
	Stage > 1	6	7

[a] SSM = superficial spreading melanoma; NM = nodular melanoma; UM = unclassified melanoma; LMM = lentigo maligna melanoma. Based on Gallagher et al. (1986b)

Table 4. Possible selection bias in case–control studies: effect of selection on control series (comparison of controls with age-matched subjects in census data of the same population; Western Canada Melanoma Study)

Parity in ever-married woman		
Parity	Interviewed controls %	Census data
0	15	15
1, 2	43	45
3, 4	29	26
5 +	13	14
Socio-economic group in males		
Group	Interviewed controls %	Census data
1	10	11
2	8	9
3	23	21
4	31	32
5	28	28

Based on Gallagher et al. (1986b)

2.10 Associations with Recorded Sun Exposure

2.10.1 Vacations

A simple measure of irregular, relatively intense exposure for people living in low sun environments is whether they take holidays in sunny places. In Canada, melanoma risk increased linearly with the frequency of such sunny holidays (Elwood et al. 1985b). We also assessed the time spent during vacations on activities such as swimming, sunbathing and beach time activities, which also showed an increasing risk with increasing exposure. In Denmark, the association was similar with a relative risk of 1.7 for those taking holidays in very sunny places (Osterlind et al. 1988a). These associations are supported by the results of smaller studies on similar populations, and by less direct evidence such as the association seen in Sweden between melanoma rates by county and the proportion of residents of the county who have a passport for foreign travel (Eklund and Malec 1978). In the Australian studies, the same technique cannot be used, as vacation travel is likely to put people at a lower rather than a higher level of sunshine.

2.10.2 Recreation

The second main measure of intermittent exposure is that of recreational activities, which are likely to give greater sun exposure than one's normal day to day life. In the

Canadian studies we defined a group of activities on the basis that a swimsuit or light clothing would be worn, such as sunbathing, swimming, or boating in warm places. There was a clear and significant positive association with such recreational activities, the relative risk rising to 2 (Elwood et al. 1985b).

The other studies, rather than attempting to measure total outdoor exposure through recreation, have reported on individual activities. The Danish study shows significant, positive increases in risk with frequency of sunbathing, and with participation in boating and in swimming (Osterlind et al. 1988a). In the Australian studies, such data are likely to be the best estimator of intermittent exposure, given that general sun exposure is high. In the Western Australia study, significant increased risks were seen with involvement in boating and fishing in the summer, and small increased risks were seen with swimming in the summer, sunbathing, and a measure of the proportion of total exposure given by such recreations in the summer, although these latter effects were not significant (Holman et al. 1986b). However, in the Queensland study there was no association seen between melanoma risk and the amount of time spent in beach activities at various periods during life. The amount of sun hours accumulated by beach activities was very substantial (Green et al. 1985a, 1986).

In addition to these associations with summer recreational exposure, the Danish study showed a moderately increased risk associated with skiing in the winter, and the Canadian study showed a similar association although the numbers were small and the effect non-significant. Winter recreational exposure would be expected to be limited to the more clearly exposed sites, and it would be interesting to know whether the Danish result was site-specific; however, unfortunately, it was not controlled for socio-economic status, and may well weaken if such a control has been used.

There is thus very good evidence of positive associations between melanoma risk and intermittent sun exposure achieved through vacation or recreational activities (Table 5). The measurements which show up strongest in the different studies tend to be different, but such differences would be expected for the different climatic conditions characteristic of the different places in which studies have been done (Elwood et al. 1987). In northerly, relatively less sunny countries such as Canada and

Table 5. Melanoma risks associated with particular activities (significantly increased risks in major studies) (Elwood et al. 1985b; Holman et al. 1986b; Osterlind et al. 1988a)

Activity	Place	Risk
Swimsuit-type activities	Canada	1.7
Sunny vacations	Canada	1.5
Boating	W. Australia	2.4
Fishing	W. Australia	2.7
Sunbathing	Denmark	1.9
Boating	Denmark	1.7
Swimming	Denmark	1.5
Sunny vacations	Denmark	1.7

Denmark, one can clearly see the effect of sunny vacations, whereas in Australia with much higher levels of general sunshine, the added effect of specific recreational exposure can be seen.

2.10.3 Occupational Exposure

In contrast to such activities, regular outdoor activities through occupation are likely to give us the best measure of regular sun exposure. According to the intermittent exposure hypothesis, these effects should be in contrast to those of intermittent exposure.

In the Canadian study, we asked questions about every job held for a period of more than 6 months, and for each job we asked questions about how much activity was regularly undertaken outdoors in summer and in winter, and what type of clothing was normally worn. A measure of occupational exposure in the summer was then constructed. There were more occupational data available for the men, and more exposure to such occupations. In men, those with the highest amount of outdoor exposure, equivalent to 32 h or more per week, had only 50% of the melanoma risk of subjects with minimum outdoor exposure; this effect was significant (Elwood et al. 1985b). In the Danish study, the results have not been presented in so much detail, but men who reported outdoor employment have a relative risk of 0.7, this being most pronounced in men who had at least 10 years of outdoor employment starting at a young age (Osterlind et al. 1988a). In the Western Australia study, hours of outdoor occupation showed a similar relationship, with a risk of 0.5 in the highest exposure category (Holman et al. 1986b). However, the Queensland study reported no association with outdoor exposure (Green et al. 1985a).

The Canadian data on occupation, however, are even more interesting. A more detailed look at the dose distribution by occupation shows a particularly interesting pattern, besides the decreased risk at high exposures there is a significant excess risk in subjects who have a modest amount of occupational exposure (Fig. 3). A subject can achieve a moderate amount of occupational exposure in principle either by having a job which regularly involves a small proportion of time outdoors, or by having a predominantly outdoor job for a short period. We explored the data to see which of these possibilities held, and found that most subjects in this category of exposure had achieved that by having a short period of outdoor employment, often early in life as a short-term job, or seasonal employment while being a student or before settling down to regular, predominantly indoor employment. The data on

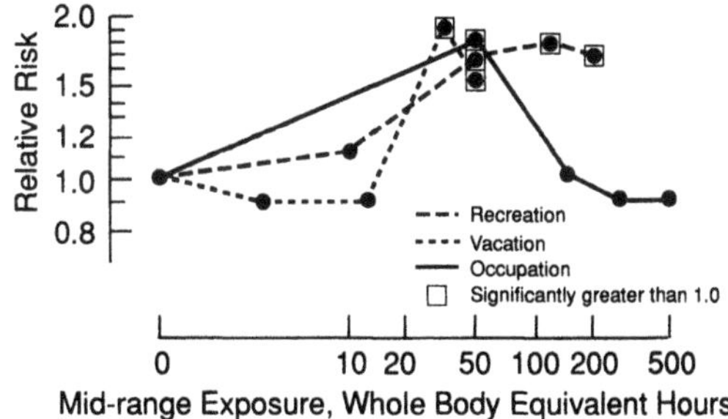

Fig. 3. Summary of the results of the Western Canada Melanoma Study, showing the relative risk of melanoma in relationship to occupational, vacation and recreational sun exposures.(Elwood et al. 1985b)

occupational exposure in Canada, therefore, provide particularly good evidence for the intermittent exposure hypothesis. An interesting complementary result is from the prospective study of Paffenbarger et al. (1978) showing that Harvard graduates who had short periods of outdoor work as students had an increased risk of melanoma in later life.

2.11 Body Site-Specific Associations

The associations so far have been between exposure of the whole body and the frequency of melanoma on any body site. In the Canadian study, analysis has not revealed any particular associations between exposure of certain body sites and the occurrence of melanoma on those body sites. However, this has been taken further in the Western Australia study which shows some interesting results; a strong association was seen between the risk of melanoma of the trunk and the type of bathing suit usually worn (Holman et al. 1986b). These data, showing effects of variation in clothing exposure in a very high sun exposure situation, are amongst the most direct evidence of a local relationship between sun exposure and local melanoma occurrence. Also, in Western Australia, the increased risk of melanoma associated with occupational exposure on particular body sites was assessed in regard to whether those sites were usually covered, sometimes exposed, or usually exposed, and these results show that the highest risk is on the intermittently exposed body sites, which is of course consistent with some of the general site-specific data.

2.12 Dose Response Relationships

Thus the relationship between melanoma risk and the dose of ultraviolet radiation received is complex and seems likely to vary by the intermittency of the dose itself, and very likely by the age at which the dose is received and the host characteristics of the subject. The Canadian study is the only one in which there has been some estimate made of the total dosages received from these different activities, by which we are able to make a rather crude comparison of the effects of the different types of exposure (Fig. 3). This shows that exposures in excess of about 20 units (1 unit = exposure of the whole body for 1 h in summer) are associated with an increase in exposure, and shows that the high observed risk with moderate levels of occupation is equivalent in dose terms to the excess seen with recreational and vacation activities. In the Canadian data there are substantial numbers of subjects who receive very high doses through regular occupation, and we do not have subjects who receive equivalently high dosages through other mechanisms.

On the basis of this empirical evidence, the dose response relationship is nonlinear. Relatively moderate, total amounts of sun exposure if achieved by intermittent, relatively severe exposure are clearly associated with an increased risk. In contrast, considerably larger total doses of sun exposure, but given by relatively constant occupational exposure over a long period of time, are not associated with any increased risk, and in fact the evidence is strong that they are associated with a decreased risk of melanoma.

It is therefore difficult to produce a unified dose response curve for the whole relationship. Intermittent and constant exposure may be intrinsically different, with conflicting effects, so that the risk for an individual depends on the balance between these two exposures. Alternatively, the relationship of melanoma risk to total solar ultraviolet dose may be a curve, risk rising from low to moderate total exposures, if achieved by intermittent exposure, and then falling again with the much higher total exposures which are achieved by chronic exposure. Armstrong (1988) suggests that a further upward trend may also apply.

But the quantification of this relationship is still largely very approximate. At what point does exposure become large or regular enough to be classified as chronic rather than intermittent? At the present stage this is an unanswerable question. The break point between the detrimental effects of so-called intermittent exposure and the beneficial effects of so-called constant exposure is likely to vary between individuals in relationship to host factors related to sun sensitivity, and may also have a great deal to do with the general background level of sun exposure characteristic of that individual.

These questions are not merely theoretical as they have implications for individual and public education. The complex dose response relationship suggests that for some individuals a reduction in total dose of sun exposure will actually increase melanoma risk. However, we do not have any evidence for a limitation of total risk on a population basis if we compare populations which are relatively homogeneous in terms of host factor characteristics. Although total sun exposure levels are so high in places like Queensland or Hawaii that theoretically those who indulged in regular recreational exposure may be getting into a constant dose situation, there is no evidence that there is any tailing off of the rapid increase in melanoma incidence with decreasing latitude or any limitation of the secular increases. Thus, all the evidence at a population level suggests that a lowering of total exposure will be associated with a lowering in total melanoma incidence, and therefore education campaigns and health promotion activities attempting to decrease total solar exposure have an appropriate goal.

2.13 Age at Exposure

Information on critical age of exposure comes from migrant studies and from the case–control studies. The Western Australia study shows that those who move to Australia from Britain and arrive before the age of 10 have melanoma rates similar to native Australians (Holman and Armstrong 1984b; Fig. 4). These migrant data suggest that the critical age is around 10–19, as arrival after age 19 confers no further benefit.

Some evidence from analytical studies supports this. In a recent study in England for example, we asked identical questions about sunburn history at ages 8–12, at ages 18–22, at the age corresponding to around 20 years before diagnosis, and in the 5 years before diagnosis (Elwood et al. 1990). Most other studies of sunburn have only asked questions about one or other of these time periods, or a general question on lifetime history. We found positive relationships with sunburn for all of these, but the question on sunburn at 12 years of age had the strongest risk and the only risk

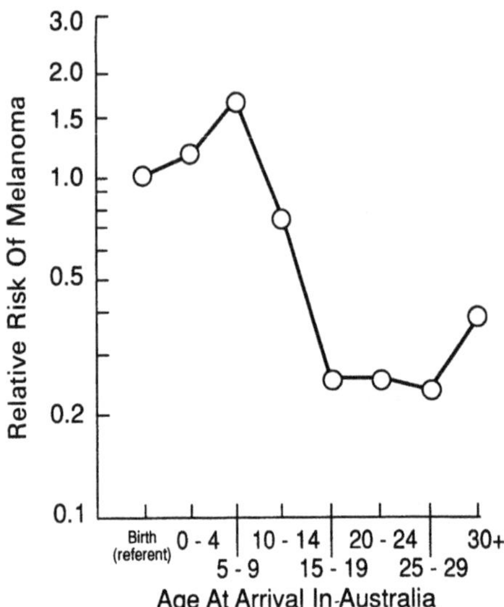

Fig. 4. Risk of superficial spreading melanoma in migrants from the UK to Australia, compared to the risk in native Australians, controlled for ethnicity. The risks for the three groups arriving at ages 20–24 and later are significantly reduced. (Holman and Armstrong 1984b)

which was individually statistically significant. The Danish study also shows the strongest associations with sunburn being at earlier ages (Osterlind et al. 1988a). Similar results come from the Nurses' Health Study in the USA, where a study of 130 melanoma patients and 300 controls showed that sunburn and residence nearer the equator at ages 15–20 were more strongly related to the risk of melanoma occurring at ages 38–65 than were such exposures later, after 30 years of age (Weinstock et al. 1989).

In the western Canada study we attempted to look at the effects of sun exposure at different times in life, but were not able to make any firm conclusions primarily because recorded sun exposure for an individual tended to be somewhat similar throughout life, i.e. the individuals with relatively high sun exposure tended to have that characteristic at all the different periods, thus making it difficult to separate the effects of differences in sun exposure at different ages. The Western Australia study was more successful in this. The associations with sunbathing and the type of bathing suit worn by women, and the association with average annual hours of bright sunlight at the place of residence, were stronger for exposures in the age range 15–24 or 10–24 than later (Holman et al. 1986b).

We may add to this the descriptive data for melanoma incidence which show clearly that melanoma of all sites except the head and neck shows a clear birth cohort-related increase (Elwood and Lee 1975), which is consistent with a change in a major causal factor in relatively early life. A different pattern might be expected if the major causal influences in melanoma occurred in a relatively short period before clinical diagnosis.

These early influences of sun exposure seem likely to be related to the natural history of acquired naevi, whose frequency rises in puberty peaking in late adolescence and early adult life, and then diminishes in the late 20s and subsequently (Armstrong and English 1988). There is good evidence that the number of naevi in young adults is related to previous sun exposure, so that sun exposure increases the frequency of such naevi, and also may stimulate their dysplastic development (Armstrong et al. 1986).

2.14 Late Effects

There is, however, evidence of a late-stage promotional effect of sunlight in the development of melanoma. The strongest evidence for this is that naevi excised during the summer in Australia show more mitotic activity than naevi excised at other times, showing a short-term effect of solar exposure on naevus development (Armstrong et al. 1984). There is also evidence for a seasonal variation in the diagnosis of melanoma, with higher rates of diagnosis in the summer, although this evidence is somewhat inconsistent and clearly it is difficult to disentangle a true incidence effect from an effect on recognition and diagnosis. Also difficult to interpret are studies of short-term annual variations in melanoma incidence, related to sunspot activity or particularly sunny summers a few years earlier. The former analyses now appear to be based on faulty logic, as years of peak sunspot activity, although linked to increased solar emission of ultraviolet radiation, are not associated with higher ground-level fluxes of ultraviolet radiation, so they do not correspond to higher solar exposure levels (Elwood 1989b). The association seen in British data between melanoma incidence and particularly sunny summers 2 years previously still holds, but has not been consistently seen in other areas (Swerdlow 1979).

It appears then that there are likely to be at least two and perhaps three time periods relevant to the relationship of sun exposure and melanoma. The migrant data suggest that the age range 10–14 is critical, and this period may relate to the initial development and occurrence of benign naevi which subsequently appear as risk markers and precursors of melanoma. There is evidence also that exposures slightly later in the range 10–24 years are important, and such exposures may relate to the interaction of further solar exposure and the existence of acquired naevi. Finally, there is some limited evidence of a late stage presumably promotional or immunological effect on the further transformation of naevi and occurrence of melanoma (Armstrong and English 1990).

2.15 Host Factors

The next issue is the interaction of sun exposure with host characteristics. We have now extremely good evidence on a large number of host characteristics which are strongly related to the risk of melanoma, including the frequency of benign, acquired naevi, the frequency and density of freckles, the natural pigmentation of the skin, hair, and eyes (which are of course strongly interrelated), and the skin type, or the propensity with which skin burns on exposure to unaccustomed sun, which is inversely related to the ability to develop a suntan (Armstrong and English 1990). Some of these factors like pigmentation are clearly genetic, others such as the number of naevi, density of freckles, and perhaps skin type are likely to represent a combination of genetic factors and the influences of early sun exposure, as there is evidence that early sun exposure is related to the development of naevi. Particularly high risks are seen in subjects with abnormal (dysplastic) naevi and a family history of melanoma (Elder 1988; Tucker and Bale 1988): intensive surveillance of such subjects may be valuable, allowing early diagnosis of new melanomas (e.g. Rigel et al. 1989). A gene for melanoma/dysplastic naevi has been found on chromosome × 1p in US families (Bale et al. 1989), but not in Dutch families (van Haeringen et al. 1989).

All the major studies have looked at sun exposure with and without adjustment for these host characteristics, and have found that the effects of external sun exposure and these host factors are generally independent. Thus, we would predict that for any individual the absolute risks associated with different categories of sun exposure will be much higher for the more susceptible subjects with several of the risk markers for melanoma. We explored in the western Canada melanoma study whether those with lower degrees of sun sensitivity, that is subjects who tanned readily and burnt rarely and had fewer freckles, appeared to be resistant to the detrimental effects of intermittent sun exposure, but within the limitations of the data there was no evidence that this was so, and the relationship with intermittent exposure appeared to be the same for all subjects (Elwood 1992). A new analysis of the New York study, comparing 289 melanoma patients with 527 other patients at the same skin clinic suggests that the effects of sunburn, outdoor occupational exposure, recreational exposure, and mixed indoor/outdoor sun exposure were greater for subjects with poor tanning ability; indeed, sunburn and a mixed indoor–outdoor pattern of occupational exposure were protective in good tanners (Dubin et al. 1989). A recent study from the US nurses' cohort also shows modification of the effects of sun exposure by tanning ability (Weinstock et al. 1991). More work on this topic is needed.

Several studies have information on the joint effect of two or more factors, and show for example that the joint effect of high numbers of naevi and high density of freckles appears to be multiplicative, with the relative risk in those who have both these characteristics being approximately equal to the multiple of the relative risks of each characteristic alone. Such a multiplicative relationship appears to be the normal interaction of two or more factors in the genesis of cancer, and is shown by many other situations of two exposure factors relating to tumour; for example in smoking and alcohol consumption in the aetiology of upper respiratory and gastrointestinal tract cancer, and smoking and age in the association with lung cancer. There is some evidence that sun exposure and individual risk factors also interact multiplicatively; thus, the Western Australia data suggest a multiplicative relationship between the number of naevi and recreational sun exposure in the production of superficial spreading melanoma (Holman et al. 1986b). So, the relative risk associated with sun exposure in subjects of high and low susceptibility to melanoma is the same, which of course means that the absolute risk is greater in those with high susceptibility. The issue of whether educational campaigns and individual surveillance programmes should be aimed at all members of a population or at high-risk subgroups defined by such host characteristics depends on the risks and benefits of such intervention and the costs involved. Relatively simple risk indicators for melanoma used individually, such as the number of acquired naevi, can readily identify 5–10% of a source population which will include 50% or more of melanomas, and the use of more complex multi-factor scoring systems, although obviously not as easy to employ, will give a somewhat better discrimination (English and Armstrong 1988; Elwood 1989a).

Thus, the information reviewed from the major studies suggests that these associations with intermittent sun exposure, and the protective effect of chronic regular exposure, are likely to apply to all subjects, with the size of the effects and the absolute risks being also determined by host characteristics. However, the major studies apply to Australia, Canada, and Denmark; all countries in which the predominant population is light-skinned, light-haired, and has a high proportion of subjects who are extremely sun-sensitive. Most of the other studies are on similar populations, such as the studies in Glasgow, studies in England, several studies in the United States, in other Scandinavian studies, and other studies in Australia and New Zealand.

We noted earlier that within Australia and North America for example, melanoma incidence rises with proximity to the equator. Within Europe this association does not hold, and melanoma rates in Scandinavia and Scotland for example are considerably higher than those in Italy, Greece, or Spain (Jensen and Osterlind 1988; Jensen et al. 1988). This aberration in trend is almost certainly due to the different host characteristics of these populations. Within a defined population there still is usually a latitude gradient, there being higher melanoma rates in the south and in the north of England and in the south and in the north of Norway for example (Jensen et al. 1988). However, Scottish rates appear to be higher than English rates, and this seems likely to be due to the higher proportion of highly sun-sensitive, typically red-haired and freckled individuals.

Further work to find out whether the associations with sun exposure and other variables seen in northern European origin populations apply to more pigmented populations is required. Consistent results are shown by a study in northern Italy (Zanetti et al. 1988). The ideal would be to use multi-centre studies using consistent clinical and epidemiological methodology to obtain data from different populations which would allow the assessment of risk factors from melanoma within those populations, and also enable comparisons to be made between the populations. This type of study is being organised through the International Agency for Research in Cancer.

2.16 The Mechanism of the Association

Let us now consider how this complex relationship between sun exposure and melanoma risk is mediated. Epidemiologists put great stress on the separation of direct effects (e.g. sun exposure causes melanoma) from indirect effects, referred to as confounders (people with a lot of sun exposure have other characteristics which increase their risk) (Elwood 1988). General expectations, and some direct evidence, suggest that individuals who adopt life-styles with high levels of intermittent sun exposure are likely to be those who tolerate sun well, and thus have host characteristics of skin type and pigmentation which are associated with a *low* risk of melanoma. The confounding between these characteristics and intermittent sun exposure is negative; controlling for skin type and pigmentation will increase the relationship between intermittent exposure and melanoma.

The other characteristics associated with melanoma are an increased number of naevi, number of dysplastic naevi, or density of freckling. These are positively associated with melanoma risk and also positively associated with intermittent sun exposure; studies show, for instance, that naevus density is highest in association with intermittent sun exposure. The most likely explanations of this are that either sun exposure is a causal factor for naevi and freckles, and is also independently a causal factor for melanoma; or that it causes naevi and freckles which in turn cause melanoma. The Western Australia data show that sun exposure and naevus count have largely independent and multiplicative effects, and therefore the more likely model is that intermittent exposure is a causal factor both for naevi and melanoma (Holman et al. 1986b).

A further major confounding factor is socio-economic status. Analysis both in terms of intermittent sun exposure and chronic sun exposure within the Canadian data shows that the association seen with socio-economic status is mediated by intermittent sun exposure; it is the opportunity for intermittent sun exposure given by affluent social circumstances which explains the association (Gallagher et al. 1987).

So we cannot explain away the associations with intermittent exposure by other characteristics associated with it. What are the possibilities for the mechanism? Intermittent sun exposure results in repeated episodes of melanocyte stimulation. It also produces new tanning reactions repeatedly, with the tan disappearing between exposures, and therefore a considerable amount of immediate tanning reaction in comparison to the amount of delayed tanning. Thirdly, it results in acute effects such as sunburn. Within one or all of these must lie the biological explanation for the melanoma production. Chronic exposure as we have seen is associated with a reduction in melanoma risk. Chronic exposure results in an episode of melanocyte stimulation resulting in immediate but then delayed tanning, and the maintenance of a suntan and higher levels of melanin in the skin. It also results in skin thickening and other changes. One clear candidate for the causative mechanism of melanoma would seem intuitively to be the repeated melanocyte stimulation related to intermittent sun exposure, and experimental evidence which could test this possible mechanism would be of great interest.

2.17 Sunburn

The relationship of melanoma with sunburn is a particularly interesting one from both theoretical and practical points of view. Sunburn is an indicator of sun damage, and is related to two major factors – the amount of intermittent exposure, and the sensitivity of the skin to such exposure. Both intermittent exposure and measures of sensitivity such as skin type are strongly associated with melanoma risk, and in addition sunburn itself is clearly associated with melanoma risk. We can explore these relationships to see which of these associations are primary and which are indirect. The joint effect of sunburn history and sunburn sensitivity, and also by contrast, the joint effect of suntan history and ability to tan easily, have been explored in both the Canadian and Australian studies, with almost identical results. In both studies, the largest and most detailed of the ones we have available, the associations seen with sunburn history disappeared when corrected for sunburn sensitivity (Elwood et al. 1985a; Holman et al. 1986b). Thus, it was the tendency to sunburn which was related to melanoma risk, not the sunburn itself. However, in the Queensland study, the association with sunburn remained after control for skin type, and the same has been seen in the Danish study and in several other studies (MacKie and Aitchison 1982; Green et al. 1985a; Osterlind et al. 1988a).

There are two main potential mechanisms of the association. One is that sunburn is merely an indicator of an acute dose in sensitive individuals. The combination of acute exposure and high sensitivity gives melanoma; it also independently gives sunburn. The second model is that sunburn itself is important in the etiology of melanoma, and that the damage related to sunburn is part of the causal chain for melanoma. If the second possibility, i.e. sunburn is in the causal chain, is true, we would expect to see a site-specific relationship between the occurrence of sunburn and the occurrence of melanoma on the same body site subsequently. There is no evidence that this is so, although it has been looked at in a number of studies, and this therefore argues against the direct hypothesis. Random error in the documentation of sunburn by site will of course make such a specific association more difficult to demonstrate.

Does the distinction matter? For an individual, avoidance of sunburn could be achieved either by reducing intermittent exposure, which all the evidence suggests would reduce melanoma risk, or decreasing sensitivity, through the use of protection or through the maintenance of a tan throughout the year. It is these latter options which are untested and for which we do not know the results. We have no information on whether reducing sensitivity to sunburn in 'artificial' ways will reduce melanoma risk. For example, a suntan could be maintained all year-round by use of sunbeds; this increases the total dose of ultraviolet exposure, and the evidence we have would suggest in general that this is likely to increase melanoma risk. Even more difficult is the issue of protection by sunblockers. What happens to melanoma risk if an individual avoids sunburn, not by reducing intermittent exposure, but by using skin protection against the acute effects of sun exposure, and maintaining their current level of intermittent exposure or even increasing it? Does this reduce melanoma risk or not? The answer relates to a number of issues; is the action spectrum for the production of sunburn the same as the action spectrum for the production of melanoma? Because if it is not, avoidance of sunburn by wearing specific blockers may not avoid the radiation related to melanoma. What are the effects of sunblockers on the transmission of solar radiation, with respect to this currently unknown action spectrum for melanoma? Are there any other effects of sunblockers and sun protection?

3 The Ozone Layer Problem and Melanoma

3.1 The Ozone Layer Problem

The epidemiological features of melanoma make it clear that the prediction of the effects of a decrease in atmospheric ozone on melanoma will not be simple. Whereas for other types of skin cancer we can assume rather simple models, such as the risk of skin cancer being proportional to total lifetime dose of solar ultraviolet radiation, we can make no such prediction for melanoma. The available evidence on melanoma in humans shows that the quantitative relationship between sun exposure and melanoma risk is complicated. The risk will depend not only on the dose of sun exposure received, but the regularity and timing of that dose, and the pigmentation characteristic of the person, which in turn are influenced by past exposure to the sun.

The highest melanoma risk occurs in individuals with low levels of chronic exposure but high levels of intermittent intense exposure; in fact, the typical exposure pattern of the affluent indoor worker in a developed country who can afford a holiday or recreational activities involving considerable sun exposure (Elwood et al. 1985b). Compared to such people, the risk in those who spend more time outdoors regularly, and so have a higher total dose of UVR, is lower. This may be due to the detrimental effects of UVR being compensated by a protective effect of chronic exposure, which may be mediated through tanning, skin thickening, or other mechanisms.

Thus, it is impossible to say for an individual what the effect of increasing the level of ambient UVR will be. For most individuals an increase in dose will move them towards the peak of increased risk, but for some an increase may decrease their risk by moving their total dose from peak levels towards those high levels which are related to protection. If we move from consideration of the individual to the consideration of population groups, the issue becomes clearer. The easiest way to assess the effects of different levels of UVR exposure on a population is to compare populations who are similar in terms of genetic and pigmentation characteristics and life-style, who live in places with differing amounts of exposure to UVR, that is, at different latitudes. We have already seen the regular trend in the incidence and mortality of melanoma, with higher rates closer to the equator (Elwood et al. 1974; Elwood and Lee 1975; Lee 1982b). There is no indication that any peak is reached; there is no point at which populations of similar pigmentation with higher exposures show a lower rather than a higher melanoma risk. Thus, the simplest way to estimate the effects of an increase in total UVR is to measure the change in incidence or death rate from melanoma which is associated with a given difference in latitude, and obtain from estimates or measurements the difference in UVR related to that latitude change. Publications using this approach in North America go back as far as 1974.

3.2 Estimates Based on Latitude and Calculated UV Levels

In 1974 we analysed death rates from melanoma in the continental United States and Canada (Elwood et al. 1974), and showed that the death rate varied dramatically with latitude. Between latitude 50°N (several Canadian provinces) to about 25°N (Florida and Texas), mortality rates more than doubled (Fig. 1). The correlation coefficient between melanoma mortality rate and latitude was very high, 0.78 for males and 0.72 for females. In other words, the melanoma death rate in any particular area can be predicted quite accurately from the latitude of that area without any other information. A variation in latitude of 2°, which is equivalent to 138 miles, carries with it a change in death rates from melanoma of about 10%.

In that study we worked with two colleagues, Drs. Mo and Green of the Interdisciplinary Centre for Aeronomy and Other Atmospheric Sciences at the University of Florida, who had developed a calculation to estimate the annual UVR flux at erythema-producing wavelengths, from information on latitude, and meteorological data on cloud cover. This calculated index of UVR exposure was very strongly correlated with latitude (correlation coefficient of 0.89), so naturally melanoma mortality rates were strongly related to this index also. The data show that a 10% increase in received UVR dosage would be expected to give an increase of 3.7–4.5% in the death rate from melanoma at a latitude of 50°, and 6.8–10.3% at a latitude of 30° (Table 6). These data showed somewhat higher values for males than for females.

Two years later, Fears, Scotto, and Schneiderman of the National Cancer Institute performed a rather similar analysis for which they had data on the incidence as well as the mortality from melanoma, and related these measurements to latitude and to a calculated measure of UVR which, as with our measure, was strongly related to latitude (Fears et al. 1976). Their data cover a slightly narrower range of latitude, and they calculated that a 10% increase in UVR would cause an increase in melanoma mortality of 7 to 12%, the higher figure applying to the more southerly latitudes which already have higher rates. In simple terms, the idea of a 5 to 10% increase in mortality for a 10% increase in UVR flux is reasonable.

Incidence rates vary more rapidly with latitude than do mortality rates, and therefore they predicted that a 10% increase in UVR would be likely to give a 14–24% increase in the incidence of melanoma. This is probably because in areas where melanoma is very common it is recognised earlier and treated earlier, and therefore a lower proportion of diagnosed patients die from the disease. Therefore, those who live in a high-risk area in terms of incidence benefit from better survival, and so the difference in mortality is less marked.

3.3 Estimates Using Measured UV Levels

In a further paper, Fears et al. (1977), instead of using calculated indices of ultraviolet radiation, used measurements from Robertson–Berger meters, although the data were available only for four areas. They used a power model, by which the calculated percentage changes are not dependent upon the initial latitude; the various mathematical models used in these studies are reviewed in Elwood (1989b). These calculations showed considerably higher effects, with an estimated 25% increase in incidence for a 10% increase in solar ultraviolet radiation.

More recent work has been more sophisticated, although the overall results are not greatly different from the earlier work (Table 6). Estimates of the impact on malignant melanoma are given in the National Research Council (NRC) 1979 document (p. 100), suggesting percentage increases in melanoma incidence of 22% and in mortality of 14% for a 10% reduction in ozone; which other sources would suggest relates to a 20% increase in UVR flux. The 1982 NRC report does not add anything beyond that given in 1979, merely commenting that the evidence for a direct relationship between ultraviolet radiation and melanoma was actually weaker

Table 6. Estimates of percentage increase in frequency of melanoma with a 10% increase in solar UV radiation. All based on USA and Canada; both sexes (simple average of sex-specific results); white populations (Elwood 1989c)

Reference	UV data	Model	50° lat		30° lat		Notes[a]
			Incidence	Mortality	Incidence	Mortality	
Elwood et al. (1974)	Calculated index	Linear		4.5		6.8	1, 2
		Exponential		3.7		10.3	
Fears et al. (1976)	Calculated index	Exponential	14.0	7.0	23.5	12.0	3, 4, 5
Fears et al. (1977)	RB meter	Power	25.0		25.0		6, 7
Scotto and Fears (1987)	RB meter	Power: T, L crude	5.5		5.5		8–12
		adjusted	3.5		3.5		
		Power: H, U crude	9.0		9.0		
		adjusted	5.5		5.5		
		Power: total crude	6.7		6.7		
		adjusted	4.2		4.2		
Pitcher (1987)	Calculated from satellite data	Power: annual		3.2		3.2	13–16
		peak		7.0		7.0	
		Exponential: annual		2.1		4.5	
		peak		5.8		8.2	

TNCS = Third National Cancer Survey, 1969–1971; SEER = Surveillance Epidemiology and End Results Program. Reproduced from Elwood (1989c)
[a] Notes: 1, Mortality data USA and Canada 1950–1979 by state/province; 58 areas. 2, UV radiation by calculation of erythema-weighted index. 3, Incidence data TNC5, 9 areas; mortality USA by state. 4, UV radiation by calculation of erythema-weighted index. 5, Calculation based on latitude equivalent to change in UV radiation. 6, Incidence data TNCS, for 4 areas. 7, UV measurements from Robertson-Berger meters, 1974. 8, Incidence data SEER, for 7 areas. 9, UV measurements from Robertson-Berger meters, 1978–1981. 10, Results presented separately by site: H, U: head and neck and upper limb; T, L: trunk and lower limb. 11, Crude results take account of age only; Adjusted results are controlled for ethnic origin, hair or skin colour, suntan lotion use, and hours spent outdoors. 12, Total, for comparison, based on 67%TL, 33%HU tumours. 13, Mortality data by US county 1950–1979. 14, UV data erythema-weighted estimate from NASA, including satellite ozone column measurements. 15, Estimates for changes in mean annual dose, and for change in peak doses (clear day in June). 16, Estimates using DNA action spectrum also made; 1–8% higher than those shown.

with the advent of more conflicting evidence (National Research Council 1979, 1982).

Scotto and Fears (1987) used annual ultraviolet counts from Robertson-Berger meters in seven areas of the USA, and data on melanoma from incidence registries. They fitted a power model and presented analyses by sex and by body site of the melanoma divided into trunk and lower limb, versus head and neck and upper limb. They obtained data on co-variates including ethnic origin, pigmentation characteristics, hours outdoors during weekdays and during weekends, and use of sunscreens, suntan lotion, and protective clothing, from telephone interviews of at least 500 households in each area. However, the method uses data only for the general population, and does not involve more direct analysis which would need such data on the melanoma patients also. Their results predict greater increases for females than for males, unlike the earlier work. The overall effects for a 10% increase in UVR are a 5.5% increase for trunk and lower limb tumours and a 9% increase for head and upper limb tumours, averaged over the two sexes. Adjustment for the various co-variates reduces the predicted increases, to a 3.5% increase for trunk and lower limb tumours, and 5.5% for head and neck and upper limb. The adjustment for natural pigmentation and ethnic origin is appropriate, as previous models by failing to take this into account may be attributing to differences in ultraviolet radiation levels, differences in melanoma experience which is actually due to genetic or pigmentation characteristics. Whether one should adjust for behavioural characteristics such as time outdoors or use of sun protection, is much more debatable. We are interested in the effect of a change in natural ultraviolet radiation levels on the human population, and if that human population changes its behaviour in response to the change in atmospheric conditions, it would seem more logical to let such behavioural changes affect the calculated prediction, rather than calculate a theoretical predicted effect on the assumption that no behaviour changes will result. Adjustment for socio-economic status might be useful but has not been done.

Further work has been done by Pitcher and Longstreth (1991). They have been working with melanoma mortality data over a 30-year period, rather like the first papers in the series, and for ultraviolet measurements are using a calculation of ultraviolet flux which is based on NASA satellite data, including measurements of ozone concentrations under high atmospheric conditions. The models fitted are complex, as they are fitted for the two sexes, for three different places covering a range of latitudes, and separately for changes in the annual UV flux and changes in the peak levels under clear summer conditions. They again find larger effects for males than for females, and find a larger effect when using the peak measurements than when using the annual measurements. The overall estimates of the percentage increase in melanoma mortality are associated with a 5% decrease in ozone level, on the assumption that this is roughly equivalent to a 10% increase in solar UV radiation, range from 2.1 to 7 at 50° latitude, and from 4.5 to 8.2 at 30° latitude.

The summary of the calculations made so far in North America (Table 6) shows considerable variability, but the highest and now rather extreme estimates are those of Fears et al. (1976, 1977), which were made using incidence data on just a few areas in the USA, and have produced estimates considerably higher than those given by any other method. The studies using mortality data are all reasonably consistent, and would suggest that a 10% increase in solar UV radiation is related to an increase in melanoma mortality somewhere between 2 and 10%. This figure is not inconsistent with the most recent data from Scotto and Fears (1987) on incidence, which range from a 3.5 to 9% increase in incidence for a 10% increase in solar ultraviolet radiation depending on the models used and the adjustments made.

3.4 Time Trend in Melanoma in Relation to Changes in Ultraviolet Radiation

If melanoma rates were otherwise stable over time, this predicted increase would give us the future trend. However, as we have seen melanoma rates are far from stable, and have been increasing rapidly over recent decades. Can we say that the trend already seen is due to ozone depletion and a consequent increase in UV radiation?

Data on ultraviolet radiation is available from Robertson-Berger (RB) meters at eight centres in the United States, ranging from 30° to 37° latitude, from 1974 to 1985 (Scotto et al. 1988). For all stations, there was a slight *downward* trend in solar UVR; the overall change was a decrease of 0.7% per year from 1974 to 1975. The RB meter is relatively insensitive to the shorter UVB wavelengths in comparison to the erythema action spectrum, and therefore a change in the RB meter reading produced by changes in ozone will be likely to be less than the change in the erythemal dose. The inconsistency between the secular changes seen in these data and other evidence of a small *decrease* in the stratospheric ozone over the northern hemisphere may be because the Robertson-Berger meters are located at airports affected by local conditions. Even so, the available data suggest that the large increase in melanoma which has been observed cannot be explained by an increase in ground-level UVR. If the trends are the effects of UVR, they must be due to an increase in personal exposure due to differences in activity and clothing habits, and as pointed out above, several facets of the increase, particularly the site specificity support this.

3.5 Prediction of Future Trends in Melanoma

This makes the prediction of future increases in melanoma very difficult. The changes in melanoma rate which have been observed over the last decades have been very large and very rapid. If indeed they are due to behavioural changes, these behavioural changes must have had dramatic effects on the amount of UVR received, and its influence on melanoma. Measurements made using personal dosimeters have suggested that a holiday involving sunbathing at a southern European location can double the total amount of solar ultraviolet radiation received by a predominantly indoor living individual in a northern European country; such results come from Great Britain, Holland, Norway, and Sweden (Diffey et al. 1982; Bernhardt and Matthes 1987; Slaper and van der Leun 1987; Schothorst et al. 1987b; Kivisakk 1987). This doubling of ultraviolet radiation is of course a much bigger effect than most predictions of ozone layer effects.

The 'worst' case scenario for a prediction of melanoma risk in the future might be that the increases which we have seen over the last few decades will continue, and to these we would add the predicted effects of an increase in UVR due to ozone layer effects. The result is a suggestion of a further rapid increase in melanoma which will make it before long one of the most common adult cancers. However, the added effect of ozone layer changes in the equation is fairly small. Further, large increases in melanoma incidence are predicted, with or without inclusion of an ozone depletion factor (Roush et al. 1985). Such a prediction was made some 10 years ago in the United States and has been confirmed by the actual changes seen (Lee 1985).

The evidence suggests that the factors which may be able to prevent that further increase are changes in behaviour; in our attitudes to sun exposure, our habits of sun exposure, and our use of protective practices and materials. Efforts at public education have been pursued most vigorously in Queensland, Australia, which has the highest rates in the world, and there is evidence over recent decades that the death rate from melanoma has stabilised (McLeod 1988). This, however, appears to be mainly due to improvements in early diagnosis and therapy, and the incidence rate of melanoma in Queensland appears to be still rising despite its already very high level, although the available data are several years old (Armstrong et al. 1982).

3.6 Latency of Effect

The possible time relationship between exposure to ultraviolet radiation and the occurrence of melanoma has been discussed earlier, and evidence for both long- and short-term effects exists.

The evidence is strongly suggestive of some or all of the ultraviolet effect having a long-term basis; thus, the sun behaviour and sun exposure of teenagers and young adults today will determine their melanoma risk over the next 20 or 40 years. Thus, the beneficial effects of changes in behaviour, and the detrimental effect of changes in the ozone layer, will be seen in future decades rather than instantaneously.

4 Artificial Sources of Ultraviolet Radiation and Melanoma

4.1 Artificial Sources of Ultraviolet Radiation

So far we have considered the associations of human skin cancers with natural sun exposure. The work on non-melanoma cancers in animals suggests that the effects of sun exposure can be explained by an association primarily with ultraviolet radiation in the B range (280–320 nm) and that the action spectrum for experimental carcinogenesis is generally similar to that for erythema. Thus questions of the carcinogenicity of a light source are often approached by assessing the ultraviolet output in minimal erythemal dosages which weight the spectral distribution in accordance with an experimentally derived erythemal action spectrum. This may well be reasonable for non-melanoma skin cancers, although there may be further effects of UVA, UVC, or chemical carcinogens, or immunological effects, in the human situation (Rundel 1983). For melanomas, no such action spectrum is available for lack of an animal model; as shown above, the epidemiological data show that sun exposure has different effects on human non-melanoma and melanoma tumours. This makes the assessment of the role of other light sources in melanoma much more difficult. The newer animal models mentioned earlier may assist in this respect.

4.2 UVR Used in Skin Therapy

An important use in the treatment of psoriasis; this is often treated by 'PUVA' (Parrish et al. 1974); orally administered 8-methoxy-psoralen and long-wave ultraviolet radiation (UVA, 320–400 nm). In a large prospective study from the USA, Stern et al. (1984) found that the risk of cutaneous squamous cell carcinoma developing at least 22 months after first exposure to PUVA was 12.8 times higher in patients exposed to a high dose of UVA than in those exposed to a low dose. This was after adjustment for exposures to previous ionising radiation and topical tar preparations. No substantial dose-related increase was noted for basal cell carcinoma. In contrast, European epidemiological surveys found no increase in skin cancer which could be directly correlated with PUVA treatment (Elwood et al. 1989). Gibbs et al. (1986) suggest that the difference in the results from the USA and Europe could reflect differences in therapeutic strategy, and cite evidence from mouse experiments in which it has been shown that the total cumulative dose of UVA necessary to produce cancer lesions was less in animals exposed to smaller daily doses than larger daily doses. This could explain why the rigid, cautious and extended USA regime appeared to produce more skin cancers than the flexible, aggressive and rapid European regime. There have been no epidemiological studies showing an association between PUVA treatment and malignant melanoma. There was no increase in melanoma seen in a 10-year study of 1380 patients with psoriasis treated with PUVA in the United States, although significant increases in colon and central nervous system tumours were seen (Stern and Lange 1988).

4.3 Fluorescent Lighting

Fluorescent lighting has been in widespread use for over 40 years. In 1982 Beral et al. published a case–control study of malignant melanoma from Sydney, Australia, with 274 female cases and 549 controls, undertaken primarily to assess the effects of oral contraceptives. A statistically significant association between the risk of melanoma and length of occupational exposure to fluorescent lighting emerged. The relative risk was highest for office workers and was 4.3 for an exposure longer than 20 years. A significant association was also found for a set of 27 male cases and 35 controls. Taking into account other factors known to be associated with melanoma, like skin colour, hair colour, number of naevi and outdoor exposure did not diminish the overall association with fluorescent light. However, the association was strongest for lesions of the trunk, which is inconsistent with a simple direct effect. This observation, arising without an a priori hypothesis, led to further epidemiological and spectral studies, which have been reviewed in detail (Elwood 1986; Maxwell and Elwood 1986; Muel et al. 1987). Although the average 'erythemal' UV output was small compared to sunlight, the output of some widely used tubes in the short wavelength part of the UVB range could be substantial and exceed that of sunlight; the question of chemical emissions was also raised (Maxwell and Elwood 1983).

Several further epidemiological studies have been reported. A postal questionnaire study in England showed no significant association (Sorahan and Grimley 1985). In Western Australia, 337 melanoma cases and 349 controls who had taken part in the major case–control study described earlier were interviewed again, by telephone. No overall association was found for melanoma in total, but a positive association between lentigo melanoma of the head, neck and upper limbs and exposure to undiffused lights in small rooms was observed. Less plausibly an association between unclassified melanoma and undiffused lighting was seen, weaker if restricted to exposure in small rooms (English et al. 1985).

Interestingly, two studies, in the USA and UK, showed a similar discrepancy in relation to the method of assessment. In New York (Dubin et al. 1986), and in England (Elwood et al. 1986), positive associations were seen in data from interviews but no association was seen in data from postal questionnaires. Normally in epidemiology, interviews are the most accurate method, but the possibility of recall bias giving a spurious positive result exists. Thus, while the weight of evidence is currently against there being a causal effect of fluorescent light, it cannot be totally dismissed.

4.4 Sunbeds

It is a common, though false, impression in affluent white societies that a good suntan equates with good health and well-being. This has led to a large increase in the use of suntanning equipment. The old type low pressure mercury arc sunlamps which emitted mainly UVB (280–320 nm) radiation have been generally condemned as much for their short-term effects (erythema and burning) as for any long-term consequences of their use. They have been largely replaced by sunbeds relying for their UV source on fluorescent lights which emit mainly UVA (320–400 nm) radiation. Type 1 lamps use barium silicate phosphor giving 95% UVA emissions; type 2 use a strontium borate phosphor and produce 99% UVA (Walter et al. 1990). The newer sunbeds produce a greater irradiance of UVA relative to UVB. As it is the UVB component of sunlight which is primarily responsible for the production of non-melanoma skin cancers, it might seem reasonable to assume that these new sunbeds are safer.

The action spectrum for human skin carcinogenesis is obviously not known, but in keeping with most work on photocarcinogenesis it can be assumed that the action spectrum for UV erythema in human skin is an adequate approximation of this (van der Leun 1984; Sterenborg and van der Leun 1987). Parrish et al. (1982) note that the melanogenic effectiveness of UVR increases with decreasing wavelength in the region between 400 and 300 nm. In view of this a larger exposure or greater intensity is required with the newer, longer wavelength sunbeds to produce a tan (Diffey 1986). Furthermore, it has been shown, at least in subjects who are genetically poor tanners, that the erythemally effective (and presumably carcinogenic) UVR dose is virtually identical to the melanogenetic action spectrum within the 300–400 nm region (Parrish et al. 1982). Thus, there is no reason to assume that getting a tan with the new type lamps is any safer.

Although UVA light does not appear to be a powerful carcinogen alone, there has been concern that it might enhance the development of tumours in people already susceptible. Staberg et al. (1983) showed that when mice were irradiated with UVA in the doses found in human solaria, there was no increase in tumour rate. However, when the mice were first irradiated with UVA and UVB (sunlight) followed by UVA alone for 2–6 months, the tumour rate was increased two-fold over those just exposed to UVA plus UVB. This could have been due to UVB contamination of the UVA source. However, when the mice were irradiated by a UVA source filtered through glass to eliminate UVB less than 320 nm the tumour-enhancing effect was maintained, albeit at a lower level. Extrapolating from this to the human situation it can be assumed that many sunbed users have already been exposed to sunshine as part of their 'sun-worshipping' tendencies and are therefore susceptible to any tumour-enhancing effect of UVA. However, Beck-Thomsen et al. (1988) found the opposite: pre-treatment for 4 weeks with UVA resulted in significantly delayed turnover production from broad spectrum UVA and UVB in mice, suggesting an anti-tumour effect of UVA.

A possible mechanism for this tumour-enhancing effect has been described. Hersey et al. (1983) observed immunological changes in the skin of subjects who used a sunbed for 12 half-hour sessions during a 2-week period. They suggested that the changes noted could be expected to impair host defences against the induction of skin tumours.

The earlier case–control studies provide little information about the use of sunlamps or sunbeds, as the numbers of subjects exposed were very low. The first study to show an effect was that of Swerdlow et al. (1988) who compared 180 melanoma patients in Scotland with hospital controls. Ultraviolet lamps or sunbeds had been used by 21% of the cases compared to 8% of the controls, giving a relative risk of 2.9, rising to 9.1 in those whose use started at least 5 years earlier. However, in Denmark, although exposure was similar, no association was seen: sunlamps had been used by 45% of melanoma cases and 42% of the controls; sunbeds by 14% of cases and 18% of controls (Osterlind et al. 1988a).

Walter et al. (1990) compared 583 melanoma patients diagnosed in 1984–1986 with 608 community controls in southern Ontario, Canada, finding use of sunbeds or sunlamps in 24% of male and 28% of female cases, compared to 15 and 21% of male and female controls. The increased risk of 1.88 for males, 1.45 for females, increased with duration and amount of use, persisted when skin colour, naevi, skin reaction to sun, and socio-economic status were controlled.

So two out of three recent studies show an effect of artificial UV sources on melanoma risk. Whether artificial and solar UV radiation, dose for dose, have the same effects, is unknown. Schothorst et al. (1987a) estimated the mutagenic effects per erythemal dose for UVR from different sources, and found that UVC, UVB, and solar irradiation had similar mutation induction per minimal erythemal dose but that UVA had a higher ratio. If confirmed, this would suggest that UVA 'tanning' may be more dangerous than sunlight.

Thus, no firm conclusion can be reached. It may be reasonable to assume that modern suntanning equipment has similar effects to natural sun exposure, and so the risk will depend on whether subjects use it as a substitute for, or a supplement to, sun exposure. One of the features to be addressed in public education campaigns on skin cancer is the lack of any support for the widely held belief that a suntan is healthy.

Suggestions of links between sunbeds and lentigenes and freckles have also been made (MacPherson and Finlay 1986; Williams et al. 1988; Roth et al. 1989), and there are reports of quite severe effects on skin fragility (Farr et al. 1988; Rademaker and Simpson 1988).

4.5 Occupational Lighting Sources and Skin Cancer

Many other sources of artificial ultraviolet light emitting radiation of wavelengths throughout the ultraviolet spectrum are used widely in industry and elsewhere. This UVR may be produced inadvertently, as the by-product of some other process, for example in welding or foundry work, where the high temperatures attained lead to the emission of ultraviolet radiation. It may also be produced deliberately in order to utilise the particular properties of UVR, for example the germicidal properties of UVC produced by low pressure mercury arc lamps (253.7 nm) which are widely used in hospitals and food processing, and the dye-line copying equipment used for industrial plans and blueprints using actinic fluorescent light whose UV emissions sensitise the copy paper. A list of occupations potentially associated with UVR exposure, both artificial and natural, has been compiled by the World Health Organization (1979).

The amount of UVR exposure from these artificial sources will depend upon a number of considerations including the spectral composition of the UVR, the radiant intensity of the source, the distance from the source, and the use or otherwise of protective shielding around the source or by the exposed subject.

Guidelines for limits for occupational exposure to UVR have been endorsed by the National Institute for Occupational Safety and Health in the USA (National Institute for Occupational Safety and Health 1972). This standard has been adopted as a voluntary standard by many European regulatory authorities including those in the UK and Sweden. However, these guidelines are based on acute effects of the skin and eyes, the minimal erythemal dose and the minimal photokeratotic dose. In view of the paucity of information about the action spectrum and dose response curve for carcinogenesis in man, they take no account of longer term carcinogenic risk. Lytle et al. (1987) and Passchier and van der Leun (1987) have reviewed the UVR output of different types of lighting and estimated its possible impact on non-melanoma skin cancer.

Larko and Diffey (1986) investigated the level of occupational exposure of UVR to staff in a dermatology department in Gothenberg, where UVR (both UVB and PUVA) is used for the treatment of psoriasis. There were 45 subjects working in different parts of the clinic. Over a 10-week period each wore two film badges on their lapels (one for recording UVB and UVC and one for recording UVA) . The study was carried out during the winter of 1984–1985 to minimise the influence of natural UVR on the relevant doses. The study showed that people working with phototherapy equipment were potentially at risk from exposure to UVR. UVA exposures were well within recommended occupational limits, but the doses of UVB and UVC exceeded recommended maximum permissible exposure in 18% of workers in one phototherapy area. Such monitoring is rare and it seems likely that excessive exposures will occur in other workplaces.

No clinical reports associating skin carcinogenesis with exposure to artificial UV sources at work have been published. We have published a pilot epidemiological study and looked at this potential problem (Elwood et al. 1986). In a case–control study of 83 patients with melanoma and an age–sex matched control group, and in the context of a lifetime occupational history, subjects were asked if they had ever worked with any particular or unusual light source, such as vacuum or discharge lamps, insecticidal or germicidal lamps or welding equipment. Twenty-one of the 83 patients compared to 11 of the controls reported having had exposure to one or more such sources: relative risk 2.2, 95% confidence limits, 1.0–4.9. This is being studied further.

While no definite occupational risks for melanoma are accepted, several occupational associations have been described. Several reports show apparently increased risks in high technology industries (Lee 1982a, b; Wright et al. 1983; Vagero et al. 1985; Gallagher et al. 1986a; Armstrong and English 1988; DeGuire et al. 1988). These associations may be due to socio-economic rather than occupational effects; increased risks are seen in professional and managerial groups in general. Several studies show increased risks in chemical engineers, which may suggest a more immediate cause. Increases in teachers and in garden and nursery workers may relate to sun exposure (Gallagher et al. 1986a). A large study of ocular melanoma showed a relative risk of 10.9 for a history of ever working as a welder (Tucker et al. 1985).

5 Conclusions

Sun exposure is the only clearly established external cause of human melanoma. It is the major cause of melanoma in white-skinned populations; programmes to modify sun exposure and hence to reduce melanoma risk are now an accepted public health priority, and the first results of such interventions are beginning to appear. The commonest types in white populations, superficial spreading and nodular melanoma, show a complex dose–response relationship to sun exposure and are particularly increased by intermittent exposure of the skin; long-term, constant sun exposure may even decrease the risk. Lentigo maligna melanoma, like non-melanoma skin cancer, may have a different dose–response relationship. The causes of other types of melanoma, melanoma of non-cutaneous sites, and melanoma in non-white populations have not been studied extensively and factors other than sun exposure may predominate. The relevant component of sunlight is ultraviolet radiation, but the action spectrum for human melanoma is unknown. While some animal models for the production of melanoma-like tumours by ultraviolet radiation have been developed, action spectra and quantitative studies have not yet been done and their relevance to the human situation will have to be demonstrated.

Exposure to some artificial sources of ultraviolet radiation also increases melanoma risk.

The main influences on individual melanoma risk are genetically determined pigmentation characteristics, and personal behaviour in regard to sun exposure. Atmospheric ozone depletion, unless countered by behavioural changes, is likely to increase melanoma risk, although the size of the projected increases is modest compared to the very large difference in risk seen between individuals with different patterns of sun exposure. Typical estimates are that a 5% decrease in ozone is likely to result in a 10% increase in ground-level ultraviolet radiation and a 5–10% increase in melanoma incidence.

Addendum

This chapter was prepared in 1990, and although brief references have been added to some later work, the considerable volume of new literature on melanoma epidemiology cannot be fully reviewed. Several important reviews have appeared recently or are in press (Koh 1991; Armstrong and English 1990; Lee 1992), and the most detailed of these is a monograph on the carcinogenic effects of ultraviolet radiation covering material up to the end of 1991 (International Agency for Research on Cancer 1992). Amongst individual studies of particular interest are studies of migrants (Mack and Floderus 1991; Khlat et al. 1992), studies showing a recent reversal of the long continued increase in incidence and mortality rates in the USA (Scotto et al. 1991; Roush et al. 1992), and studies of the modification of the effects of sun exposure by tanning ability (Weinstock et al. 1991). A related issue is that the epidemiology of non-melanoma skin cancers, which appeared to be a simple function of total solar exposure in the older literature, is now being subjected to analytical studies and is showing itself to be not simple, with intermittent exposure also playing a role (Kricker et al. 1991a, b). Attention is moving from the study of the aetiology of melanoma to its control, and information on strategies for control through public health programmes (Marks and Hill 1992; Elwood and Galsgow 1992) and evidence for the modification of sun exposure (Theobald et al. 1991) have been presented. The question of early diagnosis and screening for melanoma is an important and controversial issue which is receiving increasing attention (Koh et al. 1989, 1990; Elwood 1991; Elwood and Koh 1992; Mackie and Hole 1992; National Institutes of Health 1992). This review does not cover the advances made in other fields such as experimental carcinogenesis and genetics.

References

Arlett CF, Cole J (1989) Photosensitive human syndromes and cellular defects in DNA repair. In: Russell-Jones R, Wigley T (eds) Ozone depletion. John Wiley & Sons, Chichester, pp 147–160

Armstrong BK (1988) Epidemiology of malignant melanoma: intermittent or total accumulated exposure to the sun? J Dermatol Surg Oncol 14: 835–849

Armstrong BK, English DR (1988) The epidemiology of acquired melanocytic naevi and their relationship to malignant melanoma. In: Elwood JM (ed) Melanoma and naevi, pigment cell, vol 9. Karger, Basel, pp 27–47

Armstrong BK, English DR (1990) Cutaneous malignant melanoma. In: Schottenfeld D, Fraumeni JF (eds) Cancer epidemiology and prevention, 2nd edn. Saunders, Philadelphia (in press)
Armstrong B, Holman C, Ford J, Woodings T (1982) Trends in melanoma incidence and mortality in Australia. In: Magnus (ed) Trends in cancer incidence. Causes and practical implications. Hemisphere, Washington, pp 339–417
Armstrong B, Heenan P, Caruso V, Glancy R, Holman C (1984) Seasonal variation in the junctional component of pigmented naevi. Int J Cancer 34: 441–442
Armstrong BK, de Klerk N, Holman C (1986) Etiology of common acquired melanocytic naevi: constitutional variables, sun exposure and diet. J Natl Cancer Inst 77: 329–335
Bale SJ, Dracopoli NC, Tucker MA, Clark WH Jr, Fraser MC, Stanger BZ, Green P, Donis-Keller H, Housman DE, Greene MH (1989) Mapping the gene for hereditary cutaneous malignant melanoma-dysplastic nevus to chromosome 1P. N Engl J Med 320: 1367–1372
Beattie CW, Tissot R, Amoss M (1988) Experimental models in human melanoma research: a logical perspective. Sem Oncol 15: 500–511
Beck-Thomsen N, Wulf HC, Poulsen T, Lundgren K (1988) Pretreatment with long-wave ultraviolet light inhibits ultraviolet-induced skin tumor development in hairless mice. Arch Dermatol 124: 1215–1218
Beral V, Shaw H, Evans S et al. (1982) Malignant melanoma and exposure to fluorescent lighting at work. Lancet 2: 290–293
Bernhardt JH, Matthes R (1987) Quantitative description of exposure to UV and methods of measurements. In: Passchier WF, Bosnjakovic BFM (eds) Human exposure to ultraviolet radiation: risks and regulations. Elsevier, Amsterdam, pp 201–212
Blum HF (1959) Carcinogenesis by ultraviolet light. Univ Press, Princeton
Bridges BA (1981) How important are somatic cell mutations and immune control in skin cancer? Reflections on xeroderma pigmentosum. Carcinogenesis 2: 471–472
DeGuire L, Theriault G, Iturra H, Provencher S, Cyr D, Case BW (1988) Increased incidence of malignant melanoma of the skin in workers in a telecommunications industry. Br J Ind Med 45: 824–828
Diffey BL (1986) Use of UV-A sunbeds for cosmetic tanning. Br J Dermatol 115: 67–76
Diffey BL, Larko O, Swanberck G (1982) UV-B doses received during different outdoor activities and UV-B treatment of psoriasis. Brit J Dermatol 106: 33–41
Dubin N, Moseson M, Pasternack BS (1986) Epidemiology of malignant melanoma: pigmentary traits, ultraviolet radiation, and the identification of high-risk populations. Recent Results Cancer Res 102: 56–75
Dubin N, Moseson M, Pasternack BS (1989) Sun exposure and malignant melanoma among susceptible individuals. Environ Health Perspect 81: 139–151
Eklund G, Malec E (1978) Sunlight and incidence of cutaneous malignant melanoma. Scand J Plast Reconstr Surg 12: 231–241
Elder DE (1988) Dysplastic nevus syndrome – biological significance. Sem Oncol 15: 529–540
Elwood JM (1984) Initiation and promotion actions of ultraviolet radiation on malignant melanoma. In: Borzsonyi M, Day NE, Lapis K, Yamasaki H (eds) Models, mechanisms and etiology of tumor promotion. IARC Sci Publ 56. MRC-OUP, Lyon, pp 421–525
Elwood JM (1986) Could melanoma be caused by fluorescent light? A review of relevant epidemiology. In: Gallagher RP (ed) Epidemiology of malignant melanoma. Recent results in cancer research. Springer, Berlin, Heidelberg New York, pp 127–136
Elwood JM (1988) Causal relationships in medicine. Univ Press, Oxford
Elwood JM (1989a) Epidemiology and control of melanoma in Caucasian populations and in Japan. J Invest Dermatol 92: 214S–221S
Elwood JM (1989b) The epidemiology of melanoma: its relationship to ultraviolet radiation and ozone depletion. In: Russell-Jones R, Wigley T (eds) Ozone depletion: health and environmental consequences. John Wiley & Sons, New York London, pp 69–89
Elwood JM (1989c) The epidemiology of melanoma: its relationship to ultraviolet radiation and ozone depletion. Trans Menzies Found 15: 95–107
Elwood JM (1991) Screening and early diagnosis for melanoma in Australia and New Zealand. In: Miller AB, Chamberlain J, Day NE, Hakama M, Prorok PC (eds) Cancer screening. Cambridge Univ Press, Cambridge, pp 243–256
Elwood JM (1992) Melanoma and sun exposure: contrasts between intermittent and chronic exposure. World J Surg 16: 157–166

Elwood JM, Gallagher RP (1983) Site distribution of malignant melanoma. Can Med Assoc J 128: 1400–1404
Elwood JM, Glasgow H (1992) The prevention and early detection of melanoma in New Zealand. Cancer Society of New Zealand, Wellington (in press)
Elwood JM, Hislop TG (1982) Solar radiation in the etiology of cutaneous malignant melanoma in Caucasians. Natl Cancer Inst Monogr 62: 167–171
Elwood JM, Lee JAH (1975) Recent data on the epidemiology of malignant melanoma. Sem Oncol 2: 149–154
Elwood JM, Koh HK (1992) The conceptual basis of screening and case finding in the early diagnosis of melanoma. In: Marks R, Hill D (eds) The public health approach to melanoma control; prevention and early detection. UICC, Geneva, pp 58–71
Elwood JM, Lee JAH, Walter SD, Mo T, Green AES (1974) Relationship of melanoma and other skin cancer mortality to latitude and ultraviolet radiation in the United States and Canada. Int J Epidemiol 3: 325–332
Elwood JM, Gallagher RP, Hill GB et al. (1984) Pigmentation and skin reaction to sun as risk factors for cutaneous melanoma – western Canada melanoma study. Br Med J 288: 99–102
Elwood JM, Gallagher RP, Hill GB (1985a) Sunburn, suntan and the risk of cutaneous malignant melanoma – the western Canada melanoma study. Br J Cancer 51: 543–549
Elwood JM, Gallagher RP, Hill GB (1985b) Cutaneous melanoma in relation to intermittent and constant sun exposure – the western Canada melanoma study. Int J Cancer 35: 427–433
Elwood JM, Williamson C, Stapleton PJ (1986) Malignant melanoma in relation to moles, pigmentation, and exposure to fluorescent and other lighting sources. Br J Cancer 53: 65–74
Elwood JM, Gallagher RP, Worth AJ et al. (1987) Etiological differences between subtypes of cutaneous malignant melanoma: western Canada melanoma study. J Natl Cancer Inst 78: 37–44
Elwood JM, Whitehead SM, Gallagher RP (1989) Epidemiology of human malignant skin tumours with special reference to natural and artificial ultraviolet radiation exposures. In: Conti CJ, Klein-Szanto AJP, Slaga TJ (eds) Skin tumours; experimental and clinical aspects. Carcinogenesis: a comprehensive survey, vol 11. Raven, New York, pp 55–84
Elwood JM, Whitehead SM, Davison J, Stewart M, Galt M (1990) Malignant melanoma in England: risks associated with naevi, freckles, social class, hair colour, and sunburn. Int J Epidemiol 19: 801–810
English DR, Armstrong BK (1988) Identifying people at high risk of cutaneous malignant melanoma: results from a case–control study in Western Australia. Br Med J 296: 1285–1288
English DR, Rouse IL, Xu Z et al. (1985) Cutaneous malignant melanoma and fluorescent lighting. J Natl Cancer Inst 74: 1191–1197
Farr PM, Marks JM, Diffey BL et al. (1988) Skin fragility and blistering due to use of sunbeds. Br Med J 296: 1708–1709
Fears TR, Scotto J, Schneiderman MA (1976) Skin cancer, melanoma and sunlight. Am J Publ Health 66: 461–464
Fears TR, Scotto J, Schneiderman MH (1977) Mathematical models of age and ultraviolet effects on the incidence of skin cancer among whites in the United States. Am J Epidemiol 105: 420–427
Fisher MS, Kripke ML (1977) Systemic alteration induced in mice by ultraviolet light irradiation and its relationship to ultraviolet carcinogenesis. Proc Natl Acad Sci USA 74: 1688–1692
Fisher MS, Kripke ML (1978) Further studies on the tumor-specific suppressor cells induced by ultraviolet radiation. J Immunol 121: 1139–1144
Freeman SE, Hacham H, Gange RW, Maytum DJ, Sutherland JC, Sutherland BM (1989) Wavelength dependence of pyrimidine dimer formation in DNA of human skin irradiated in situ with ultraviolet light. Proc Natl Acad Sci USA 86: 5605–5609
Gallagher RP, Elwood JM, Threlfall WJ, Band PR, Spinelli JJ (1986a) Occupation and risk of malignant melanoma. Am J Ind Med 9: 289–294
Gallagher RP, Elwood JM, Hill GP (1986b) Risk factors for cutaneous malignant melanoma – the western Canada melanoma study. Recent Results Cancer Res 102: 38–55
Gallagher RP, Elwood JM, Threlfall WJ et al. (1987) Socioeconomic status, sunlight exposure, and risk of malignant melanoma: the western Canada melanoma study. J Natl Cancer Inst 79: 647–652
Gibbs NK, Honingsmann H, Young AA (1986) Puva treatment strategies and cancer risk. Lancet 1: 150–151
Green A (1984) Sun exposure and the risk of melanoma. Aust J Dermatol 25: 99–102

Green A, Siskind V, Bain C (1985a) Sunburn and malignant melanoma. Br J Cancer 51: 393–397

Green A, MacLennan R, Siskind V (1985b) Common acquired naevi and the risk of malignant melanoma. Int J Cancer 35: 297–300

Green A, Bain C, McLennan R (1986) Risk factors for cutaneous melanoma in Queensland. Recent Results Cancer Res 102: 76–97

Hanawalt PC, Cooper PK, Ganesan A, Smith CA (1979) Sunlight, DNA repair, and skin cancer. Am Rev Biochem 48: 783–836

Hersey JLM, Bradley M, Hasic E, Haran G, Edwards A, McCarthy WH (1983) Immunological effects of solarium exposure. Lancet 1: 545–548

Holman CDJ, Armstrong BK (1984a) Cutaneous malignant melanoma and indicators of total accumulated exposure to the sun: an analysis separating histogenetic types. J Natl Cancer Inst 73: 75–82

Holman CDJ, Armstrong BK (1984b) Pigmentary traits, ethnic origin, benign nevi, and family history as risk factors for cutaneous malignant melanoma. J Natl Cancer Inst 72: 257–266

Holman CDJ, Armstrong BK, Heenan PJ (1983) A theory of the etiology and pathogenesis of human cutaneous malignant melanoma. J Natl Cancer Inst 71: 651–656

Holman CDJ, Armstrong BK, Heenan PJ et al. (1986a) The causes of malignant melanoma: results from the West Australian lions melanoma research project. Recent Results Cancer Res 102: 18–37

Holman CDJ, Armstrong BK, Heenan PJ (1986b) Relationship of cutaneous malignant melanoma to individual sunlight-exposure habits. J Natl Cancer Inst 76: 403–414

International Agency for Research on Cancer (1992) IARC monographs on the evaluation of carcinogenic risks to humans; ultraviolet radiation, vol 55. IARC, Lyon (in press)

Jensen OM, Osterlind A (1988) Host–environment interactions of malignant melanoma of the skin. In: Fortner JG, Rhoads JE (eds) Accomplishments in cancer research 1988. J B Lippincott, Philadelphia, pp 167–188

Jensen OM, Carstensen B, Glattre E, Malker B, Pukkala E, Tulinius H (1988) Atlas of cancer incidence in the Nordic countries. Nordic cancer union. Cancer Societies of Denmark, Finland, Iceland, Norway and Sweden

Khlat M, Vail A, Parkin M, Green A (1992) Mortality from melanoma in migrants to Australia: variation by age at arrival and duration of stay. Am J Epidemiol 135: 1103–1113

Kivisakk E (1987) Intentional exposure to ultraviolet radiation: risk reduction and present regulations. In: Passchier WF, Bosnjakovic BFM (eds) Human exposure to ultraviolet radiation: risks and regulations. Elsevier, Amsterdam, pp 443–454

Koh HK (1991) Cutaneous melanoma. N Engl J Med 325: 171–182

Koh HK, Lew RA, Prout MN (1989) Screening for melanoma/skin cancer: theoretic and practical considerations. J Am Acad Dermatol 20: 159–172

Koh HK, Caruso A, Gage I, Geller AC, Prout MN, White H, O'Connor K, Balash EM, Blumental G, Rex IH Jr, Wax FD, Rosenfeld TL, Gladstone GC, Shama SK, Koumans JA, Baler GR, Lew RA (1990) Evaluation of melanoma/skin cancer screening in Massachusetts. Cancer 65: 375–379

Kraemer KH, Lee MM, Scotto J (1987) Xeroderma pigmentosum: cutaneous, occular, and neurologic abnormalities in 830 published cases. Arch Dermatol 123: 241–250

Kricker A, Armstrong BK, English D, Heenan PJ, Randell PL (1991a) A case–control study of non-melanocytic skin cancer and sun exposure in Western Australia (Abstr No III, P3). Cancer Res Clin Oncol 117 (Suppl II): S75

Kricker A, Armstrong BK, English DR, Heenan PJ (1991b) Pigmentary and cutaneous risk factors for non-melanocytic skin cancer – a case–control study. Int J Cancer 48: 650–662

Kripke ML (1979) Speculations on the role of ultraviolet radiation in the development of malignant melanoma. J Natl Cancer Inst 63: 541–545

Lancet (1989) Sunlight, DNA repair, and skin cancer. Lancet 1: 1362–1363

Larko O, Diffey BL (1986) Occupational exposure to ultraviolet radiation in dermatology departments. Br J Derm 114: 479–484

Lee JAH (1982a) Melanoma. In: Scottenfeld D, Fraumeni JF (eds) Cancer epidemiology and prevention. Saunders, Philadelphia, pp 984–995

Lee JAH (1982b) Melanoma and exposure to sunlight. Epidemiol Rev 4: 110–136

Lee JAH (1985) An update on the epidemiology of malignant melanoma. Rep Environ Protect Agency (unpublished)

Lee JAH (1992) Melanoma and other cutaneous neoplasms: etiology, risk factors, epidemiology, and public health issues. Curr Opin Oncol (in press)
Lee JAH, Strickland D (1980) Malignant melanoma: social status and outdoor work. Br J Cancer 41: 757–763
Lehmann AR (1987) Cockayne's syndrome and trichothiodystrophy: defective repair without cancer. Cancer Rev 7: 82–103
Ley RD, Applegate LA, Padilla RS, Stuart TD (1989) Ultraviolet radiation-induced malignant melanoma in *Monodelphis domestica*. Photochem Photobiol 50: 1–5
Libow LF, Scheide S, DeLeo VA (1988) Ultraviolet radiation acts as an independent mitogen for normal human melanocytes in culture. Pigment Cell Res 1: 397–401
Lytle CD, Hitchins VM, Beer JZ (1987) Estimation of carcinogenic risk from lamps which emit ultraviolet radiation. Int Congr Ser 744: 193–197
Mack TM, Floderus B (1991) Malignant melanoma risk by nativity, place of residence at diagnosis, and age at migration. Cancer Causes Control 2: 401–411
MacKie RM, Aitchison T (1982) Severe sunburn and subsequent risk of primary cutaneous malignant melanoma in Scotland. Br J Cancer 46: 955–960
MacKie RM, Hole D (1992) Audit of public education campaign to encourage earlier detection of malignant melanoma. Br Med J 304: 1012–1015
MacPherson TD, Finlay AY (1986) Ultraviolet A freckles: another hazard of sunbeds? Br Med J 292: 380
Marks RM, Hill DJ (1992) Melanoma control manual. UICC, Geneva
Marks R, Ponsford MW, Selwood TS, Goodman G, Mason G (1983) Non-melanotic skin cancer and solar keratoses in Victoria. Med J Aust 2: 619–622
Maxwell KJ, Elwood JM (1983) UV radiation from fluorescent lights. Lancet 2: 579
Maxwell KJ, Elwood JM (1986) Could melanoma be caused by fluorescent light? A review of relevant physics. In: Gallagher RP (ed) Epidemiology of malignant melanoma. Recent results in cancer research. Springer, Berlin, Heidelberg, New York, pp 137–138
McLeod GR (1988) Control of melanoma in high-risk populations. In: Elwood JM (ed) Naevi and melanoma: incidence, interrelationships and implications: pigment cell 9. Karger, Basel, pp 131–139
Muel B, Cesarini J-P, Elwood JM (1987) Malignant melanoma and fluorescent lighting: report of technical committee to the Comite International de l'Eclairage. CIE. In: Passchier WF, Bosnjakovic BFM (eds) Human exposure to ultraviolet radiation: risks and regulations. Elsevier, Amsterdam, pp 89–96
Muir CS, Nectoux J (1982) Time trends in malignant melanoma of the skin. In: Magnus K (ed) Trends in cancer incidence: causes and practical implications. Hemisphere, Washington, pp 365–386
National Institute for Occupational Safety and Health (ed) (1972) Criteria for a recommended standard occupational exposure to ultraviolet radiation. DHEW, Washington
National Institutes of Health (1992) Diagnosis and treatment of early melanoma. NIH Consensus Development Conference consensus statement 1992 Jan 27–29; 10(1), National Institutes of Health, Bethesda, pp 1–25
National Research Council (ed) (1979) Protection against depletion of stratospheric ozone by chlorofluorocarbons. Nat Acad Sci, Washington
National Research Council (ed) (1982) Causes and effects of stratospheric ozone reduction: an update. Nat Acad Sci, Washington
Newell GR, Sider JG, Bergfelt L, Kripke MI (1988) Incidence of cutaneous melanoma in the United States by histology with special reference to the face. Cancer Res 48: 5036–5041
Office of Population Census and Surveys (ed) (1980) Cancer statistics, survival 1971–73. Ser MB1, No 3. HMSO, London
Office of Population Census and Surveys (ed) (1987) Mortality statistics, cause 1985. Ser DH2, No 12. HMSO, London
Office of Population Census and Surveys (ed) (1988) Cancer registrations 1984. Ser MB1, No 16. HMSO, London
Osterlind A, Tucker MA, Stone BJ et al. (1988a) The Danish case–control study of cutaneous malignant melanoma. II. Importance of UV-light exposure. Int J Cancer 42: 319–324

Osterlind A, Tucker MA, Hou-Jensen K et al. (1988b) The Danish case–control study of cutaneous malignant melanoma. I. Importance of host factors. Int J Cancer 42: 200–206

Paffenbarger RS, Wing Al, Hyde RT (1978) Characteristics in youth predictive of adult-onset malignant lymphomas, melanomas and leukemias: brief communication. J Natl Cancer Inst 60: 89–92

Parrish JA, Fitzpatrick TB, Tanenbaum L, Pathak MA (1974) Photochemotherapy of psoriasis with oral methoxsalen and longwave ultraviolet light. N Engl J Med 291: 1207–1211

Parrish JA, Jaenicke KF, Anderson RR (1982) Erythema and melanogenesis action spectra of normal human skin. Photochem Photobiol 36: 187–191

Passchier WF, van der Leun JC (1987) Human exposure to ultraviolet radiation: limits for unintentional UV exposure. Int Congr Ser 744: 531–535

Pitcher HM, Longstreth JD (1991) Melanoma mortality and exposure to ultraviolet radiation: an empirical relationship. Environ Int 17: 7–21

Rademaker M, Simpson N (1988) Skin fragility and sunbeds. Br Med J 297: 358

Rigel DS, Rivers JK, Kopf AW, Friedman RJ, Vinokur AF, Heilman ER, Levenstein M (1989) Markers for increased risk for melanoma. Cancer 63: 386–389

Roth DE, Hodge SJ, Callen JP (1989) Possible ultraviolet A-induced lentigines: a side effect of chronic tanning salon usage. J Am Acad Dermatol 20: 950–954

Roush GC, Schymura MJ, Holford TR (1985) Risk for cutaneous melanoma in recent Connecticut birth cohorts. Am J Publ Health 75: 679–682

Roush GC, McKay L, Holford TR (1992) A reversal in the long-term increase in deaths attributable to malignant melanoma. Cancer 69: 1714–1720

Rundel RD (1983) Promotional effects of ultraviolet radiation on human basal and squamous cell carcinoma. Photochem Photobiol 38: 569–575

Schothorst AA, Enninga EC, Simons JW (1987a) Mutagenic effects per erythemal dose of artificial and natural sources of ultraviolet light. Int Congr Ser 744: 103–107

Schothorst A, Slaper H, Telgt D, Alhadi B, Suurmond D (1987b) Amounts of ultraviolet B (UVB) received from sunlight or artificial UV sources by various population groups in the Netherlands. In: Passchier WF, Bosnjakovic BFM (eds) Human exposure to ultraviolet radiation: risks and regulations. Elsevier, Amsterdam, pp 269–273

Scotto J, Fears TR (1987) The association of solar ultraviolet radiation and skin melanoma incidence among Caucasians in the United States. Cancer Invest 5: 275–283

Scotto J, Cotton G, Urbach F, Berger D, Fears T (1988) Biologically effective ultraviolet radiation: surface measurements in the United States, 1974 to 1985. Science 239: 762–764

Scotto J, Pitcher H, Lee JAH (1991) Indications of future decreasing trends in skin-melanoma mortality among whites in the United States. Int J Cancer 49: 490–497

Setlow RB, Woodhead AD, Grist E (1989) Animal model for ultraviolet radiation-induced melanoma: platyfish-swordtail hybrid. Proc Natl Acad Sci USA 86: 8922–8926

Slaper H, van der Leun JC (1987) Quantitative modelling of skin cancer incidence. In: Passchier WF, Bosnjakovic BFM (eds) Human exposure to ultraviolet radiation: risks and regulations. Elsevier, Amsterdam, pp 155–171

Sorahan T, Grimley RP (1985) The aetiological significance of sunlight and fluorescent lighting in malignant melanoma a case–control study. Br J Cancer 52: 765–769

Staberg B, Wulf HC, Poulson T, Klemp P, Brodhagen H (1983) Carcinogenic effect of sequential artificial sunlight and UV-A irradiation in hairless mice. Arch Dermatol 119: 641–643

Sterenborg HJ, van der Leun JC (1987) Action spectra for tumorigenesis by ultraviolet radiation. Int Congr Ser 744: 173–191

Stern RS, Lange R (1988) Cardiovascular disease, cancer, and cause of death in patients with psoriasis: 10 years prospective experience in a cohort of 1,380 patients. J Invest Dermatol 91: 197–201

Stern RS, Laird N, Melski J, Parrish JA, Fitzpatrick TB, Bleich HL (1984) Cutaneous squamous-cell carcinoma in patients treated with PUVA. N Eng J Med 310: 1156–1161

Stierner U, Rosdahl I, Augustsson A, Kagedal B (1989) UVB irradiation induces melanocyte increase in both exposed and shielded human skin. J Invest Dermatol 92: 561–564

Swerdlow AJ (1979) Incidence of malignant melanoma of the skin in England and Wales and its relationship to sunshine. Br Med J 2: 1324–1327

Swerdlow AJ, English JSC, MacKie RM, O'Doherty CJ, Hunter JAA, Clark J, Hole DJ (1988) Fluorescent lights, ultraviolet lamps, and risk of cutaneous melanoma. Br Med J 297: 647–650

Theobald T, Marks R, Hill D, Dorevitch A (1991) "Goodbye sunshine": effects of a television program about melanoma on beliefs, behavior, and melanoma thickness. J Am Acad Dermatol 25: 717–723

Tucker M, Bale SJ (1988) Clinical aspects of familial cutaneous malignant melanoma. Sem Oncol 15: 524–528

Tucker MA, Shields JA, Hartge P, Augsburger J, Hoover RN, Fraumeni JF Jr (1985) Sunlight exposure as risk factor for intraocular malignant melanoma. N Engl J Med 313: 789–792

Urbach F, Epstein JH, Forbes PD (1974) Ultraviolet carcinogenesis: experimental, global, and genetic aspects. In: Pathak MA, Harber LC, Seiji M, Kukita A, Fitzpatrick TB (eds) Sunlight and man: normal and abnormal photobiologic responses. Univ Press, Tokyo, pp 259–283

Vagero D, Ahlbom A, Olin R, Sahlsten S (1985) Cancer morbidity among workers in the telecommunications industry. Brit J Ind Med 42: 191–195

Van der Leun JC (1984) UV-carcinogenesis. Photochem Photobiol 39: 861–868

Van Haeringen A, Bergman W, Nelen MR et al. (1989) Exclusion of the dysplastic nevus syndrome (DNS) locus from the short arm of chromosome 1 by linkage studies in Dutch families. Genomics 5(1): 61–64

Walter SD, Marrett LD, From L, Hertzman C, Shannon HS, Roy P (1990) The association of cutaneous malignant melanoma with the use of sunbeds and sunlamps. Am J Epidemiol 131: 232–243

Weinstock MA, Colditz GA, Willett WC, Stampfer MJ, Bronstein BR, Mihm MC, Speizer FR (1989) Nonfamilial cutaneous melanoma incidence in women associated with sun exposure before 20 years of age. Pediatrics 84: 199–204

Weinstock MA, Colditz GA, Willett WC, Stampfer MJ, Bronstein BR, Mihm MC Jr, Speizer FE (1991) Melanoma and the sun: the effect of swimsuits and a "healthy" tan on the risk of nonfamilial malignant melanoma in women. Am J Epidemiol 134: 462–470

Williams HC, Salisbury J, Brett J et al. (1988) Sunbed lentigines. Br Med J 296: 1097

World Health Organisation – WHO (ed) (1979) Environmental health criteria 14. Ultraviolet radiation. WHO, Geneva

Wright WE, Peters JM, Mack TM (1983) Organic chemicals and malignant melanoma. Am J Ind Med 4: 577–581

Zanetti R, Rosso S, Faggiano F, Roffino R, Colonna S, Martina G (1988) A case–control study on cutaneous malignant melanoma in the province of Torino, Italy. Rev Epidemiol Sante Publique 36: 309–317

2 Maintaining Health of Farm Animals in Adverse Environments

E. SIMENSEN

1 Introduction

Experience shows that climatic factors influence the incidence and severity of animal diseases, and adverse weather conditions are often cited as the cause of outbreaks of infectious disease. There is therefore a need to protect farm animals from extreme weather conditions in order to ensure a good state of health.

The effect of hot and cold weather on performance in livestock has been extensively studied under controlled conditions. A solid bank of knowledge, primarily based on production and physiological data, provides a rational basis for general recommendations on temperature requirements for livestock (see Hahn 1981). In contrast, the effect of weather on animal health is not as well documented. Thus, insufficient research prevents a detailed description of the thermal requirements of livestock in relation to health criteria.

This chapter represents a review of the influence of climatic factors, primarily air temperature and humidity, on animal health. Special emphasis is given to field and experimental observations on animal diseases, together with the influence of air temperature and humidity on the animal's resistance to disease. The influence of climatic factors on the survival of airborne microorganisms is also discussed. The review does not deal with parasitic or nutritionally based diseases.

1.1 The Multifactorial Aetiology of Infectious Diseases

Contact between pathogenic microorganisms and animals may lead directly to disease. This applies, for example, in the case of virulent pathogens such as foot and mouth disease virus, and rabies virus, where environmental conditions have little effect on the incidence of disease once animals have become infected. However, disease is more often caused by less virulent, and more chronic and insidious, pathogens, which are present either alone or in a mixed flora in the environment in which the animals are kept. Infection may be regarded as a natural state, and disease results when there is a disturbance in the equilibrium between the host, pathogen and environment. Ways in which climatic factors may influence animal diseases are schematically represented in Fig. 1.

1.1.1 Host Defence

The external environment may affect the local and/or general defence of the host in many ways. For example, climatic factors may exert a direct, adverse effect on the

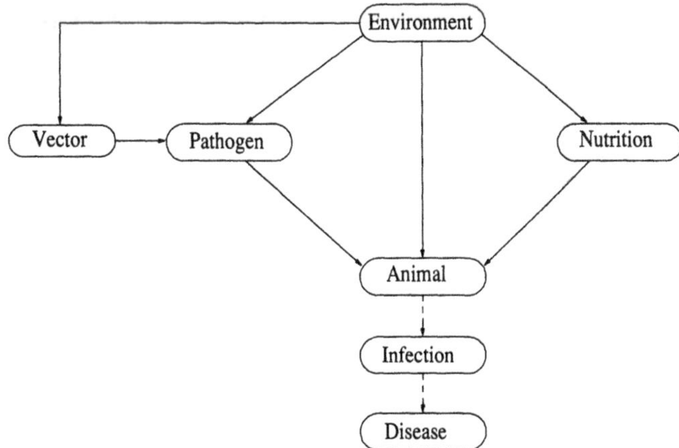

Fig. 1. A schematic presentation of the interplay between the external environment, pathogen and animal, which influences resistance to infectious disease. Disturbance in equilibrium results in infection and disease

skin and the mucous membranes of the respiratory tract. These tissues are in contact with the environment, and direct injury to them facilitate entry of pathogenic microorganisms through these important natural barriers. Sunburn and frostbite are examples of such adverse effects. Climatic factors such as heat and cold may also act as physiological stress factors which affect the specific and non-specific responses of the body to infection.

1.1.2 Pathogen Survival

Climatic factors may affect dispersal, spread and survival of pathogenic microorganisms in the environment. This is also true for arthropod vectors such as mosquitos and ticks (Smith 1970; Ferguson and Branagan 1972). The density of the animal population is an important factor determining the concentration of pathogens in the environment. Population density can be influenced by weather conditions, as animals respond to heat and cold by typical changes in behaviour. For example, in cold weather they tend to huddle together. This behaviour results in increased population density, which in turn involves an increased risk of the spread of airborne infections.

1.1.3 Nutritional Effects

Deviations from normal weather conditions during the growing season can influence plant growth both quantitatively and qualitatively. Thermal factors thus can result in nutritional disorders of animals which are dependent on plant growth. Such disorders can in turn reduce resistance to disease (Ferguson and Branagan 1972). As an example, avitaminosis A occurs widely in domestic ruminants in climates characterized by a long and dry growth season and associated paucity of green food. The condition is known to result in impairment of resistance of epithelial membranes to infection.

The possible interactions, as outlined in Fig. 1, clearly indicate that there is no specific cause of disease. The influence of climatic factors on the incidence, distribution and severity of infectious disease should always be viewed against this background.

2 Field Observations

Field data on the relationship between weather and disease originate from empirical observations by herdsmen or veterinarians, or from epidemiological studies in which data have been statistically evaluated in relation to various weather parameters.

2.1 Cold Conditions

Several diseases in various animal species, particularly respiratory disease, have been related to cold conditions.

Bovine respiratory disease is a disease complex in which major lesions are localized in the respiratory tract. In cattle, several authors have reported a correlation between season and respiratory disease incidence, with peaks during the winter season, particularly late autumn, early winter and late winter (see review by Dennis 1986).

A sudden drop in environmental temperature is often followed by outbreaks of acute respiratory disease within 24 to 72 h. Wiseman et al. (1976) reported that this association between temperature and respiratory disease was most obvious in housed calves. However, sudden onset of cold and wet weather conditions was also observed to precipitate the disease in single, suckled calves that had never been housed. It has been suggested that cold causes latent respiratory infections to flare-up and that the tendency of animals to huddle together to counteract the effects of cold temperatures facilitates the spread of infectious agents (Webster 1981).

Outbreaks of pneumonia in housed calves have been related to high relative humidity, which is usually observed during mild winter weather (Jones and Webster 1981).

The incidence of pneumonia in feedlot cattle was closely associated with the daily temperature range, i.e. the difference between maximum and minimum temperature, and the concentration of dust particles (2.0 to 3.3 μm in diameter; MacVean et al. 1986). There appeared to be a lag time period of 15 days in the autumn and 10 days in the spring between exposure and onset of disease.

Workers studying specific infectious agents causing respiratory disease in cattle have also found a relationship between season and disease incidence. In Western Europe, the diseases caused by bovine respiratory syncytial disease virus and parainfluenza virus III are most commonly seen during the autumn (Andresen et al. 1981; Pririe 1981). Germain et al. (1975) found that serum antibodies against parainfluenza virus III in cattle were more prevalent in winter than in summer, and that infection with the virus was most prevalent in cold weather with high humidity. Clinical manifestations of infectious bovine rhinotracheitis have also been related to sudden changes in temperature, and cold, wet weather. The highest incidence of bovine rhinotracheitis was observed during autumn and winter (Dougherty 1971).

The respiratory disease complex in pigs has also been associated with season and weather (Dennis 1986; Witte 1986). Sanford and Josephson (1981) observed that outbreaks of pneumonia caused by *Haemophilus pleuropneumonia* frequently followed cold weather, winter storms, transportation or other form of stress. Shope (1955) observed outbreaks of swine influenza among pigs on different farms following a change to cold and wet weather.

Enzootic pneumonia in housed pigs, caused by *Mycoplasma hyopneumoniae* is most common during the winter. A sudden drop in outdoor temperature has been associated with the development of acute, severe cases of pneumonia in fattening pigs housed in poorly insulated buildings (Whittlestone 1976).

Sheep pneumonia has a seasonal incidence (Kriton et al. 1976). Stress factors, including inclement weather and marked changes in weather, are considered to be important in the precipitation and severity of the disease (Blood et al. 1983).

McIlroy et al. (1989) observed a seasonal pattern of pleuritic and pneumonic lesions in sheep slaughtered in northern Ireland. The appearance of these lesions was consistent from year to year and the incidence was highest during late winter and lowest during midsummer. The prevalence of lesions increased dramatically 2 months after periods with rainy and windy weather combined with low temperatures.

Shearing increases the vulnerability of sheep to climatic factors. Deaths among newly shorn sheep due to pneumonia have been reported after the onset of wind- and rainstorms. Also, in hot, dry climates, post-shearing deaths have been reported in connection with extreme fluctuations between high temperatures during the day and low temperatures during the night (Hugh-Jones 1976).

Calf mortality in California has been related to climatic factors (Martin et al. 1975). In winter, an increased mortality rate was significantly associated with cold, wet and windy weather.

Mastitis in dairy cows has been observed to be more prevalent during periods of inclement weather (Schalm et al. 1971). Hropot (1970) and Simensen (1974) reported a higher incidence of mastitis during periods of cold and wet summer weather.

Bovine virus diarrhoea has been reported to occur most frequently during the winter in housed and ranged animals (Blood et al. 1983). Cold winter weather also influences the clinical course of *keratoconjunctivitis in calves.* During cold weather (−20 °C) calves reared outdoors were more severely affected than animals housed indoors, although the disease was of shorter duration in animals kept outdoors (Kopecky et al. 1981).

Early reports on the epizootiology of *Newcastle disease in poultry* indicated that outbreaks occurred mainly during periods of cold weather, and tended not to appear during hot weather in enzootic areas (Brandly et al. 1946). In turkeys, outbreaks of *fowl cholera,* caused by *Pasteurella multocida,* have been related to changes from warm and dry to cold and wet weather conditions (Alberts and Graham 1948).

Erysipelas in swine has long been linked to sudden changes in weather. Sudden exposure to cold temperatures has been reported to enhance susceptibility to erysipelas (Wood 1986).

2.2 Hot Conditions

In Australia, heat stress is considered to be a major cause of *calf mortality*. Increased mortality rates have been observed during hot and dry weather years, and most of the heat-stressed calves which died had diarrhoea (Mitchell et al. 1981). It was observed that the cattle tended to gather around water troughs in hot and dry weather periods, and the risk of contaminating the water with faeces was thus increased. Greater calf mortality was also associated with hot, dry weather in California (Martin et al. 1975).

A relationship between high environmental temperatures and *mastitis in dairy cows* has not been reported, although hot weather has been associated with increased somatic cell counts in milk (Nelson et al. 1967). Another study failed to demonstrate any relationship between high environmental temperature and somatic cell counts (Roussel et al. 1969).

Erysipelas in swine has been related to sudden exposure to excessive, hot temperatures or exposure to sustained high temperature (above 30 °C; Wood 1986).

Fowl cholera in turkeys has been related to hot conditions. During periods of hot weather, an increase in the number and severity of fowl cholera outbreaks was observed (Olson et al. 1968). The relationship between weather and disease was most evident after a drop in temperature, when the incidence of fowl cholera decreased markedly. Deaths of poultry from other diseases such as colisepticemia have also been associated with periods of excessive heat (Cheville 1979).

2.3 Summary

The majority of associations between weather and disease reported in farm animals involve respiratory disease which is often related to cold stress. Change in weather to cool and humid conditions in the autumn and spring, and the daily variations that occur during these periods, appear to predispose animals more to disease than does exposure to severe cold temperatures alone. This corresponds to the view of Andrews (1964), who held that the most meaningful data relating human diseases to environment concerned sudden changes in weather rather than temperature, humidity and air velocity per se. Exposure to cold temperatures has also been related to enteric and other diseases. In addition, reports have also linked disease outbreaks with exposure to high temperatures, although these cases are few when compared to those resulting from cold exposure.

3 Experimental Observations

Animals have been exposed to a variety of stresses including hot and cold temperatures together with pathogenic microorganisms under controlled experimental conditions. Results have been reviewed by Siegel (1974), Kelley (1980, 1982), Simensen (1985), and Dennis (1986) among others.

The most significant results relating environmental temperature to the course of disease have been obtained from studies of viral infections. Many, but not all experiments, showed that cold stress, especially without prior conditioning, reduces the resistance while heat stress can increase resistance of animals to disease.

Several experiments have also demonstrated increased resistance in hot environments and increased susceptibility in cold environments to bacterial infections. However, equivocal results have been reported.

3.1 Air Temperature and Viral Infection

An early study that demonstrated a relationship between air temperature and viral disease concerned *infectious laryngotracheitis in poultry* (Hudson 1931). The mortality rate in chickens inoculated with the virus and kept at ambient temperatures of 37 to 39 °C was lower than in inoculated chickens kept at a moderate temperature. Using the same pathogen, Sinovic (1970) observed the highest mortality in birds kept at 5 °C.

Another example of the relationship between air temperature and viral infection is *transmissible gastroenteritis in pigs.* Two- to 3-month-old pigs, maintained at 30 °C and exposed to the virus, showed no signs of diarrhoea. A sudden decrease in air temperature to 4 °C, either before or after virus inoculation, induced severe disease. When the pigs were exposed to a cold ambient temperature (4 °C) for 14 days before inoculation, fewer animals showed clinical signs of disease. This finding suggests that adaptation to cold temperature had taken place in the animals (Shimizu et al. 1978).

In contrast to the above results, resistance of poultry to *Newcastle disease* was highest at low temperatures and lowest at high temperatures (Sinha et al. 1957). After infecting 6- to 7-week old chickens with aerosols containing the Newcastle disease virus, mortality was 100% at 29.5 to 32 °C, 90% at 21 to 23.5 °C, 75% at 10 to 13 °C, and 55% at 0 °C. The incubation period was shorter, and nervous signs more pronounced at the higher temperatures, whereas respiratory signs were more marked in birds at the lower temperatures.

In an experiment with *avian influenza* virus, cold stress at −1 °C had no effect on the susceptibility of turkeys to the organisms (Robinson et al. 1979).

3.2 Air Temperature and Bacterial Infection

Experiments concerning *fowl cholera* in turkeys have demonstrated temperature-dependent responses. Simensen et al. (1980) kept groups of 7- to 8-week old turkeys at 33 to 35 °C, 19 to 22 °C and 2 to 5 °C, and exposed them to *Pasteurella multocida* either by inoculation of the palatine air spaces or through the drinking water. At the highest temperatures, depression and mortality were delayed, whereas mortality was highest at the lowest temperatures. Turkeys exposed to cold for a period of 1 week before inoculation were more depressed than those maintained at moderate temperatures before inoculation.

In contrast, experiments with 12-week-old chickens have shown that exposure to environmental temperatures of 35 or 39 °C after intramuscular inoculation with *Pasteurella multocida* increased mortality and decreased survival time (Juszkiewicz 1967; Juszkiewicz et al. 1967). Cold exposure of the birds at −1 or −2 °C had no effect on the mortality rate, but appeared to reduce the development of infection. There was less growth of *P. multocida* in the liver, spleen, kidneys and lungs in the cold-stressed birds and more in the heat-stressed birds when compared to birds kept at 18 °C.

Experiments with *Escherichia coli* infections in pigs have shown temperature-related responses. Weaned pigs kept at 28 °C and orally challenged with the bacterium gained weight faster, and had less diarrhoea than inoculated pigs kept in a cold environment (Armstrong and Cline 1977). In another study, 3-week-old piglets, challenged with a pathogenic strain of *E. coli*, developed post-weaning diarrhoea at an air temperature of 15 °C, but not at 20 to 30 °C (Wathes et al. 1985).

It has been hypothesized that the effect of cold on the pathogenesis of *E. coli* infections is related to changes in the intestinal microflora. The frequency of diarrhoea, and its severity, were greater in piglets that were infected at 18 h of age and maintained at 25 °C than pigs maintained at 35 °C (Mendoza Sarmiento 1986). Further, the intestinal peristaltic activity in pigs kept at 25 °C was reduced or absent. However, reduced peristaltic activity was not shown to mediate the increased susceptibility to *E. coli*.

3.3 Air Humidity and Bacterial and Viral Infections

Experiments have demonstrated effects of ambient relative humidity (RH) on experimental infections. Some, but not all, results indicate that the effect of RH is temperature-dependent.

Relative humidity influences the severity of *rhinotracheitis in turkeys*, which is caused by *Alcaligenes faecalis*. Turkey poults were inoculated nasally, and exposed to 75 to 80% RH or 20 to 35% RH. All birds were kept at 29 to 35 °C (Slavik et al. 1981). Birds exposed to the high humidity treatment were more adversely affected than those exposed to the low humidity treatment. At the higher RHs, clinical disease was more severe, infection occurred sooner after inoculation, and the recovery time was prolonged, compared to birds maintained at the lower RHs.

An experiment was performed with *Mycoplasma synoviae* infections in chickens under different temperature and humidity conditions (Yoder et al. 1977). The organism was administered in an aerosol, and chickens were exposed to high (31 to 32 °C), medium (19 to 24 °C) or low (7 to 10 °C) air temperatures, and high (75 to 90%), medium (38 to 56%), or low (23 to 26%) relative humidities. Air-sac lesions were most extensive in birds at low temperatures, regardless of RH. At medium air temperatures, the incidence of airsacculitis was greater at low than at high RH. At high temperatures, there was a trend towards more airsacculitis at high humidity than at low humidity.

The effect of climatic stress on *bovine rhinotracheitis* was studied by Jericho and Darcel (1978). Calves were subjected to a number of constant ambient temperature and humidity combinations, and exposed to bovine herpes virus 1 in an aerosol. Although all calves became febrile, the pattern of gross and microscopic lesions in the lungs was not influenced by the climatic conditions to which the calves had been exposed.

3.4 Air Temperature and Responses in Non-Inoculated Animals

A possible effect of cold temperature on the pathogenesis of *mastitis* in dairy cows may be related to the local cooling of the udder. The effects of local cooling were studied in an experiment where two cows were held in temperature-controlled rooms for successive 5-day periods at moderate (21 to 28 °C), cold (– 16 °C), moderate (21 to 28 °C), hot (36 to 37 °C) and moderate environments. For two other cows the cold and hot sequences were reversed (Brown et al. 1977). Intramammary temperatures were 1 to 2 °C higher than deep body temperature in the hot room and 9 °C lower in the cold room. Somatic cell counts in milk showed no consistent changes during the period of heat and cold stress, and the effect of environmental temperature on intramammary infections was also inconsistent. Bacterial counts in milk in relation to heat and cold appeared to vary with the type of organism.

Ewbank (1968) found that surface cooling of the front quarters was associated with an increase in the cell count of the milk; however, this finding could not be confirmed by Holmes (1971).

Cold stress has been reported to cause *diarrhoea* in experimental animals, without any preceding challenge with pathogenic microorganisms. Kelley et al. (1982a) placed neonatal piglets either in a thermoneutral (35 °C) or cold (21 °C) environment. After 48 h, the survival rate was 62% for animals kept at 35 °C and 36% for those kept at 21 °C. Diarrhoea was more prevalent in the cold-exposed pigs.

Calf health was studied in an experimental in which newborn animals were housed in hutches for the first 3 weeks of life in a thermal environment that cycled daily between – 20 and – 8 °C, or – 30 and – 18 °C (Rawson et al. 1988). Clinical assessment of calves revealed no significant effects due to cold treatment. At necropsy, cold-housed calves had subcutaneous haemorrhage and oedema of the hindlimbs (Rawson et al. 1988). Olson et al. (1980a) also observed similar cold-induced lesions in the hind legs of severely cold-stressed, hypothermic calves.

Weaned pigs were subjected to intermittent, daily, unpredictable draught (air velocity 1 m/s and air temperature 7 °C below the environmental temperature) (Scheepens 1991). Compared to control pigs maintained at an air velocity below 0.2 m/s, the draught-exposed pigs had a decreased growth rate, increased coughing, sneezing and diarrhoea and skin lesions, and more severe pneumonic lesions.

An increase in the severity of diarrhoea under suboptimal climatic conditions, as observed by Scheepens (1991), corresponds with results from experimental *Escherichia coli* infections as previously discussed. The results are also in agreement with the findings of Dividich et al. (1982) and Curtis and Morris (1982), who found more scouring when nocturnal temperatures were reduced for weaned piglets. Pouteaux et al. (1982) found an increased passage rate of faeces when environmental temperatures were reduced.

3.5 Air Temperature and Virus Replication

Thermal stress can influence the replication of pathogenic virus in animals. After infecting newborn piglets with transmissible gastroenteritis virus, Furuuchi and

Shimizu (1976) found the highest viral levels in the tissues of animals maintained at 8 to 12 °C. Virus was not found in animals at 35 to 37 °C, while at 20 to 23 °C, virus was found only in the lymph nodes and respiratory organs. Shope (1955) showed that swine in a controlled environment could carry lungworm larvae infected with swine influenza virus without showing signs of influenza. The disease could, however, be precipitated by exposing the animals to cold temperatures.

Another study demonstrated that temperatures effects on viral replication operate at a local level rather by systemic modification of body defence systems. To illustrate, an experiment was performed with bovine herpes virus 2, a virus that causes bovine herpes mammalitis in cows (Lechtworth and Carmichael 1984). Two areas of thoracic skin were inoculated with the virus at different temperatures. In one experiment the subcutaneous temperature was 1 to 3 °C above and in another 2 to 9 °C below the rectal temperature of the animals. Lesions in cold skin appeared sooner after inoculation, were larger and deeper, contained more infectious virus, viral antigen and interferon, and lasted longer than lesions in warm skin.

3.6 Summary

The examples cited clearly demonstrate that heat and cold stress influence the pathogenesis of experimental infections from a wide variety of bacteria and viruses. However, some experiments have failed to demonstrate any effect of heat and cold on infection, and in other experiments, non-infected animals have been subjected to temperature extremes without any apparent effect on health.

Relative humidity (RH) also plays a part in susceptibility to infection, although few experiments have dealt with these relationships. Results indicate that the effect of RH is temperature-dependent.

4 Survival of Airborne Microorganisms

Weather may influence animal diseases by affecting the survival of pathogens either by acting directly on the free-living stage of the organism outside the animal or indirectly on an intermediate vector in the disease cycle. Climatic factors that influence dispersal, spread, and survival of airborne microorganisms have been reviewed (Donaldson 1978; Wathes 1987).

Many microorganisms which are pathogenic for farm animals are known to be aerially transmitted. Wind has been reported to transmit viral diseases over wide distances (Smith 1970; Gloster et al. 1982; Knight et al. 1985; Raddatz et al. 1991).

Diseases caused by airborne microorganisms mainly affect the respiratory tract, although airborne transmission of enteric and other diseases can also occur. Relative humidity and temperature often are major factors influencing the survival of airborne microorganisms.

High concentrations of airborne microorganisms are known to exacerbate respiratory diseases due to increased challenge because more organisms are deposited in the respiratory tract per unit of time (Pritchard et al. 1981). High concentrations are found particularly in indoor housing of animals due to high population density. Thus, Curtis and Drummond (1982) observed concentrations of airborne microorganisms in calf houses which were up to 1000 times greater than those in an outdoor environment.

4.1 Survival of Bacteria

In general, an increase in air temperature results in an increased death rate of airborne bacteria. For example, the half-lives of *Escherichia coli* were approximately five times shorter at 30 °C than at 15 °C and over a wide range of relative humidities (Wathes et al. 1985). Mycoplasma organisms also show an inverse relationship between survival and temperature. Airborne spores are, however, heat-resistant.

Regarding relative humidity, there appears to be a zone of instability for most microorganisms. Thus, a variety of bacteria have been shown to be most sensitive to mid-range RH. For *E. coli*, death is more rapid at low RHs (below 50%; Wathes 1987). Abrupt changes in RH can result in significant changes in viability of airborne microorganisms. Jones and Webster (1981) observed a significantly positive correlation between RH and the concentration of airborne bacteria. Further, they found that the concentration of viable organisms in a naturally ventilated calf house was five times higher during periods of mild, humid weather (maximum room temperature 14 °C and RH 80–85%) than during cold, less humid weather (minimum temperature 3 °C and RH below 60%).

The variation in concentrations of airborne microorganisms due to fluctuations in temperature and RH is also reflected in the bacterial colonization of the mucous membranes of the respiratory tract. At a constant temperature of 16 °C, bacterial populations in the nasopharynx of calves were at a minimum between 65 and 75% RH, and tended to increase at RHs outside this range (Jones and Webster 1984). It was further observed that the number of bacteria in the nasopharynx of 9-week-old calves was positively correlated with lung damage in the same animals at 16 to 18 weeks of age.

4.2 Survival of Virus

Virus survival appears to be less influenced by environmental temperature when compared to bacterial survival. However, low temperatures were found to favour the survival of influenza (Harper 1961) and Newcastle disease viruses (Songer 1967).

Airborne viruses are sensitive to RH, although the effect of RH varies by species (see review by Donaldson 1978). At constant temperatures, viruses that have a lipoprotein envelope survive best at low RH, whereas non-enveloped viruses survive best at high RH. Newcastle disease virus is most stable at low RH, whereas foot and mouth disease virus survives best at high RH, although there are differences between strains (Donaldson 1978). At an ambient temperature of 18–23 °C, some viruses were more sensitive to RH at mid-range levels (30–70% RH) than at high (80%) or low (20%) RH (Donaldson and Ferris 1976).

The aerosol stability of the infectious bovine rhinotracheitis virus has also been studied (Elzhary and Derbyshire 1979). In nasal secretions from non-infected calves, the virus survived best at 90 or 30% RH when the temperature was 6 or 32 °C, respectively.

4.3 Summary

High concentrations of airborne microorganisms exacerbate respiratory disease due to increased challenge of pathogenic microorganisms via the respiratory tract.

Airborne bacteria are generally sensitive to high environmental temperatures, while virus survival is less dependent on temperature.

Relative humidity is a major factor influencing survival of airborne bacteria and virus at constant air temperatures. High RH favours survival of many farm animal pathogens. Abrupt changes in RH can result in significant changes in viability of airborne microorganisms. Little progress has yet been made in the study of the effect of RH on virus survival at varying air temperatures.

5 Host Defence

Host defence to infectious disease can be classified into local and general types. The external surfaces of the body, such as skin and the mucous membranes of the respiratory tract, represent physical barriers that protect against the entry of pathogenic organisms. These barriers constitute the local host defence system. General host defence is provided by the specific and non-specific immune systems involved in systemic resistance to infection. Examples of the specific and non-specific defence systems include the phagocytic systems and humoral and cellular immunity. Both types of resistance are influenced by adverse environmental factors.

5.1 Local Defence

Hot and cold temperatures influence peripheral blood circulation and skin temperature. As previously discussed, surface cooling of animals induced lesions in peripheral tissues (Olson et al. 1980a; Rawson et al. 1988; Scheepens 1991), increased leucocytic cell counts in milk (Ewbank 1968), and favoured viral replication in the skin (Lechtworth and Carmichael 1984). Environmental temperature may also influence the natural microbial flora on the external surfaces of animals and disturb the host-pathogen balance (Lehner and Sasshofer 1984).

Climatic factors influence the respiratory tract epithelium and its clearance of viable organisms. It has been hypothesized that pulmonary infection may result from environment-induced impairment of lung clearance mechanisms (Jones and Webster 1984). The effects of air temperature and humidity on the defence mechanisms of the upper and lower respiratory tract have not been studied extensively.

5.1.1 Colonization and Spread of Microorganisms in the Upper Respiratory Tract

The effect of changes in climate on bacterial colonization of calf respiratory tract has been studied (Jones 1987). Calves were exposed to an aerosol-generated inoculum of *Pasteurella haemolytica* and then subjected to an abrupt climate change, from 5 °C/72% RH to 13 °C/84% RH. This change in conditions caused a transient increase in the respiratory rate in infected calves and resulted in a rapid colonization of *P. haemolytica* in the nasopharynx.

Environmental temperature and RH have been shown to influence the spread of virus in the respiratory tract in chickens inoculated with Newcastle disease virus (Baetjer et al. 1960). Following nasal inoculation, the spread of the virus at a constant temperature (22 °C) was greatest at high RH (90%). At a high temperature (29 °C), the extent of virus spread was greatest at low RH (20%).

5.1.2 Mucociliary Clearance

It has been stated that low RH impairs mucociliary action. Jericho and Magwood (1977) found that low RH (35%) at 30 °C had a drying effect on the upper respiratory tract epithelium of calves. Goblet cells were least common in calves held in hot and dry air, while calves held in dry air had the lowest polymorphonulcear cell counts of the upper respiratory tract epithelium. Calves in the dry environment also had the highest prevalence of hypochromatic cell layers and vacuolation of respiratory epithelial cells.

The study of Baetjer et al. (1960) indicated that inhalation of warm, dry air appeared to inhibit the ciliary mechanisms, thus facilitating the spread of viral infection in the respiratory tract.

Other studies have failed to demonstrate any effect of RH on respiratory tract epithelium. Humans were exposed to 70, 50, 30 and 10% RH at 23 °C (Anderson et al. 1972). No changes were observed in the mucus flow or in nasal airflow resistance at the four humidity levels, and no relationship was observed between a subject's mucus flow and susceptibility to infections of the upper respiratory tract.

Heat and cold stress can inhibit respiratory clearance mechanisms. In stress reactions, elevated sympathetic activity has the effect of diminishing the water content of respiratory tract mucus, thus increasing mucus viscosity. Glucocorticoids can also inhibit mucociliary action (Breazile 1988).

Effects of air temperature on the pathogenesis of respiratory disease have been related to pulmonary oxygen tension. When animals are exposed to cold temperatures, respiratory and pulmonary ventilation rates are reduced to minimize evaporative heat loss. This may induce a degree of hypoxia, known to reduce both tracheal mucociliary action and alveolar macrophage activity (Thomson and Glinka 1974).

A study was performed in which calves inhaled radiolabelled inert particles measuring 3.3 μm in diameter (Jones and Bull 1987). Neither mucociliary clearance rate nor alveolar deposition of particles was affected by a change in conditions from 14 °C/87% RH to 5 °C/75% RH. The authors concluded that climatic variations most likely influence respiratory disease through an effect on biocidal clearance rather than physical lung clearance.

5.2 *General Defence*

Heat and cold are classical stress factors that induce adrenal responses. The immunological and neuroendocrinological consequences of climatic stress have been discussed (Kelley 1985; Moberg 1985).

Increased levels of glucocorticoids inhibit phagocytic cell activity and suppress lymphoid cells (see Roth 1985; Claman 1987; Breazile 1988). Cellular and humoral immune responses are thus modified by adverse weather conditions, provided the conditions are severe enough to elicit an adrenal response. Some examples of heat and cold effects on the immunity of farm animals under experimental conditions are discussed in the following sections.

5.2.1 Phagocytic Systems

Stress-induced inhibition of macrophage activity can reduce clearance of inhaled bacteria found in the pulmonary alveoli. Curtis et al. (1976) showed that cold (6 °C) inhibited pulmonary clearance of non-pathogenic *Escherichia coli* in piglets. The inhibitory effect of cold stress became progressively less as the pigs grew older.

In chickens, short-term heat stress caused a significant decrease in blood leucocyte counts (Ben Nathan et al. 1976; Heller et al. 1979). Further, heat-exposed chickens were found to phagocytize *Staphylococcus aureus* more rapidly than non-treated animals (Heller et al. 1979).

5.2.2 Absorption of Colostral Immunoglobulins

In calves, exposure to high temperatures reduces the amount of immunoglobulins transferred from colostrum to blood (Stott et al. 1976). Similarly, extreme cold stress reduced maternally transferred immunity. Newborn calves were immersed in cold water until body temperature was decreased by 10 °C. This treatment delayed the onset and decreased the absorption of colostral immunoglobulins (Olson et al. 1980b). A less extreme cold exposure (1 °C for 3 days) did not significantly influence absorption of colostral immunoglobulins by calves (Olson et al. 1981; Olson and Bull 1986). Short-term cold exposures at 10 °C reduced the capability of newborn pigs to acquire colostral immunoglobulins (Blecha and Kelley 1981a). In another experiment with newborn piglets, cold stress (21 °C), sufficient to induce hypothermia, did not impair absorption of colostral immunoglobulins (Kelley et al. 1982a).

5.2.3 Circulating Antibodies

Exposure to heat and cold influences the levels of circulating antibodies. Serum IgG levels in calves were significantly lower in a cold than in a thermoneutral environment (Rafai et al. 1985). Kelley et al. (1982b) observed a reduction in serum IgG concentrations in calves exposed to 35 °C for 2 weeks, whereas levels of IgM were not influenced. Cold exposure for 2 weeks at −5 °C slightly reduced IgG levels but had no effect on IgM.

Growing chickens were immunized against a variety of antigens and later subjected to short, intermittent, high temperatures (four episodes of 42 °C over a 4-h period). The chickens responded with a reduction in humoral antibodies to the antigens used within 12 h of exposure (Thaxton and Siegel 1970).

5.2.4 Antibody Synthesis

Studies with growing chickens have shown that short, intermittent, heat stress reduces the development of humoral immunity to a variety of antigens (see e.g. Subba Rao and Glick 1970; Thaxton and Siegel 1970). However, acute, intermittent, heat exposure (42 °C) did not affect the capability of birds to synthesize antibodies (Regnier et al. 1980). When birds were exposed to 36 °C for 5 days, antibodies to sheep red blood cells were generally higher than in control birds kept at 26 °C (Regnier and Kelley 1981). A time relationship between immunization and initiation of stress has been observed. Stress at the time of antigen contact results in reduced antibody response (Gross and Siegel 1988). Severe, short stress applied to chickens within 4 days after immunization resulted in elevated antibody titers to sheep red blood cells and *Escherichia coli* (Heller et al. 1979).

Chronic cold stress generally appears to increase antibody titers in several species. In 5-week-old pigs exposed to 0 °C for 4 days, synthesis of antibodies against sheep red blood cell antigens was stimulated (Blecha and Kelley 1981b). In a study with transmissible gastroenteritis virus-infected pigs (Shimizu et al. 1978), serum-neutralizing titers to the virus were elevated in the cold-stressed animals. In chickens, the peak antibody response to sheep red blood cells was not affected by chronic cold stress lasting for 6 days (Regnier and Kelley 1981).

5.2.5 Cell-Mediated Immunity

Heat and cold stress influence cell-mediated immunity. For example, calves subjected to chronic heat stress (35 °C) for 2 weeks showed a significant reduction in tuberculin reactions (Kelley et al. 1982b). Cold exposure (− 5 °C) for 1 week resulted in an increased tuberculin response, whereas cold exposure for 2 weeks suppressed the response. Contact sensitivity reactions to dinitrofluorobenzene and phytohemagglutinine A were also suppressed after 2 weeks of cold exposure. Chronic heat and cold stress impaired expression of contact sensitivity reactions in chickens (Regnier and Kelley 1981).

Weaned pigs were exposed to intermittent, daily, unpredictable draught (Scheepens 1991). Compared to non-exposed pigs, the exposed animals showed lower skin test lymphocyte blastogenic responses after mitogenic stimulation with phytohaemagglutinin and concanavalin A.

5.3 Summary

Climatic factors modify both local and general host defence. Air temperature and RH influence colonization and spread of microorganisms in the upper respiratory tract and mucociliary clearance. Sudden changes in temperature and RH, and also low air humidity, reduce the local resistance of the respiratory tract. Cold exposure also reduces pulmonary clearance.

Several studies have demonstrated the effects of heat and cold on humoral and cellular immunity. Climatic factors have, under experimental conditions, influenced absorption of colostral immunoglobulins, circulating antibodies, antibody synthesis, and cell-mediated immunity. Many of these results were from experiments involving short-term extreme, thermal exposure.

6 General Discussion and Conclusions

6.1 Host–Parasite Relationships

Although the empirical studies discussed in Section 2 (field observations) do not provide conclusive evidence, they nevertheless clearly indicate that climatic factors are important in the aetiology of various infectious diseases. Effects can be mediated via local or general host defence, pathogen survival in the environment, nutrition, or through a combination of all these factors (Fig. 1).

Results from studies with experimental infections as discussed in Section 3, support the hypothesis that heat and cold, and also RH, modify host responses to disease. However, in many cases experimental results are equivocal, and not in agreement with results obtained from field observations. Factors, such as route of inoculation, duration of heat and cold exposure, timing of thermal stress in relation to inoculation, etc. may explain these differences.

As discussed in Section 4, air temperature and RH are often main factors influencing dispersal, spread and survival of airborne microorganisms outside the animals. Climatic factors are therefore important with respect to the extent to which the animals are challenged with pathogenic microorganisms. In Section 5, experimental results are discussed that demonstrate effects of climatic factors on local and general host defence to infection.

From the data discussed in Sections 2–5, it can be concluded that complex mechanisms underlie interactions between pathogens and host defence to infection, and that there is no single specific cause of disease.

The relative importance of the various factors involved in the host–parasite balance is dependent on the disease. In the case of respiratory disease, emphasis is placed on pathogen survival, particularly in housed animals. In the opinion of Webster (1981, 1983), the association between cool, damp conditions and respiratory disease is almost entirely a result of the increased amount of challenge from the pathogen, rather than reduction in host resistance. In the case of animals kept outdoors, it is more likely that climatic factors (heat or cold) influence host resistance by affecting immune responses. This hazard is, however, offset by a much lower microbial challenge.

6.2 Thermal Requirements in Relation to Animal Health

In a discussion of housing and management practices aimed at reducing the effect of climate on livestock, Hahn (1985) introduced the concept of "performance penalties" as a consequence of livestock being exposed to thermal environments with which they cannot cope. Hahn further discussed how performance penalties can be estimated and used as a basis for general guidelines on temperature requirements for farm animals.

The consequences in terms of reduced performance are related to impaired reproduction, lower production, or poor health. The effects of adverse climatic conditions on reproduction and production have been studied quite extensively. A solid bank of data is available about these relationships. The data can be used as a rational basis for general recommendations on temperature requirements for livestock and for housing and management decisions to avoid performance penalties (see Hugh-Jones 1989).

Less work has been done in this area with respect to animal health, although considerable progress has been made in the last few years in the efforts to unravel how environmental stresses influence an animal's resistance to disease.

Performance penalties resulting from impaired health are difficult to assess, as they are not readily quantified in terms of adverse climatic environments. The quantitative relationship is linked rather to an increasing risk that disease can occur, although other factors must be present before disease outbreaks actually take place, or an otherwise harmless disease takes a more severe course. In this case, drastic economic consequences may arise. It follows from these risk considerations that it is difficult to draw a definite boundary between an acceptable and unacceptable environment in relation to animal health.

Risk assessments can be made principally through the concept of threshold limits, which is important in developing rational environmental criteria. Threshold limits can represent discontinuities or highly variable responses in otherwise continuous biological functions, as in the case of health problems associated with adverse weather conditions (see Sect. 2). Health problems can be expressed in terms of increased mortality rates or disease incidences, or prevalences of subclinical diseases with the main effect being a reduction in overall performance. By coupling thresholds to biological functions, they can be used for statistical evaluation based on functional relationships between loss thresholds and animal and environmental factors.

Figure 2 illustrates the concept of risk associated with threshold limits for performance losses and the increased level of vulnerability regarding health problems as influenced by animal and environmental parameters. In situation A, genetics, performance level and environmental influences combine to create a low level of vulnerability. Increased performance level, as in a high-production level dairy cow, increases the vulnerability of the animal (situation B). Coupling situation B with an adverse environment can put the animal at risk for loss (situation C). Thus, a high-performance animal, in other than optimal environment, can be at risk for loss. In situation D, inherent genetic characteristics which are disadvantageous to the animal in coping with the environment, combined with a high performance level, immediately puts the animal at risk for loss of performance. In this situation, an adverse environment can increase animal vulnerability and managerial risk to unacceptable levels.

Loss threshold evaluations are illustrated by the work of Simensen and Norheim (1983), in which calf health in Norwegian dairy herds was related to growth rate. The results provided a statistical basis for health management where deviations from the norm became unacceptable. In herd health programs in several countries targets

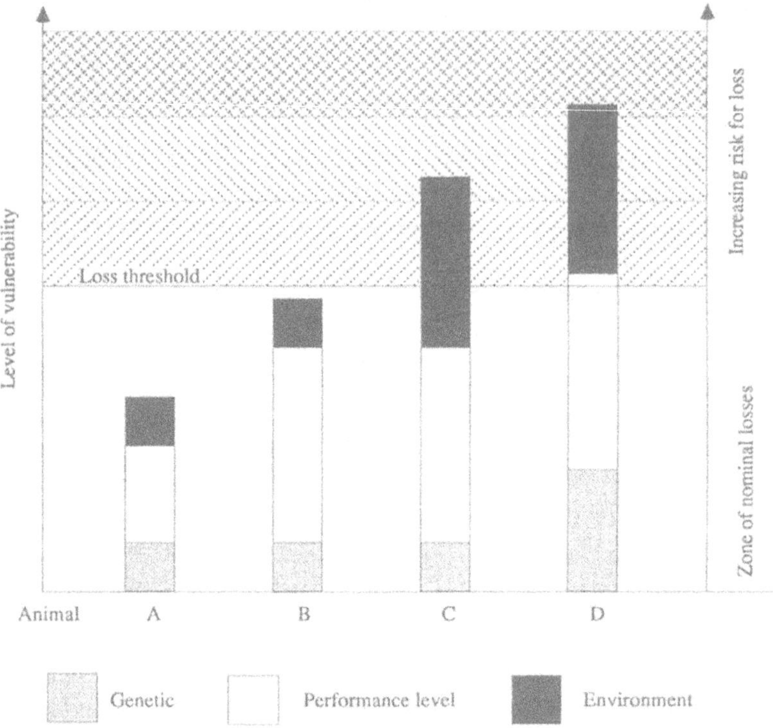

Fig. 2. Concept of risk as related to animal performance level, genetic adaptability and the environment

have been set for mortality rates, incidence rates of disease, etc. Above the pre-set threshold herd investigations are implemented to provide a basis for corrective actions.

From the data discussed in this chapter it can be concluded that climatic factors play an important role in the host–parasite relationships and that they modify animal resistance to disease.

6.3 Research needs

Although farmers and veterinarians often claim that there is an association between weather and disease, particularly change to cold and humid conditions and outbreaks of respiratory disease, there are relatively few results from epidemiological studies that document such relationships. Epidemiological research should therefore be given priority to provide more observational data from the field.

Field studies are necessary to describe the disease problem, and to evaluate statistical associations between disease and climatic factors. Properly designed epidemiological studies are well suited for revealing causal factors which are relevant for practical disease control. The attributable risk for disease associated with adverse weather conditions can be assessed by means of risk analyses, while decision or cost-benefit analyses provide data regarding the economical consequences of alternative management decisions (Fetrow et al. 1985; Thrusfield 1988). Such data are important when evaluating the need and justification for interventions (environmental modification, housing, etc.).

In Section 6.2 it is concluded that present knowledge is still fragmentary and superficial and not sufficient to provide specific criteria for thermal requirements of farm animals to ensuring optimal health. In the last few years, great advances have, however, been made in understanding the molecular basis for host resistance, particularly immune function. Growing evidence supports that there is a two-way communication between the immune and neuroendocrine systems, which allows a delicate modulation of immune responses. Several recent reviews discuss these relationships (Breazile 1988; Kelley 1988; Blalock 1989; Griffin 1989; Khansari et al. 1990).

Further information about the function of the neuroendocrine and immune systems will enable us to understand how thermal stress influences host resistance to disease. However, present knowledge is still too fragmentary and superficial to specify the thermal requirements for optimal health, or to prevent health problems associated with weather extremes.

More fundamental immunological research is needed before such knowledge can be applied to practical animal husbandry conditions.

The host–pathogen relationship, particularly with regard to weather and respiratory disease, is another area in which more research is needed. Local resistance of the respiratory tract is important in determining the cause of respiratory disease. Few reports have dealt with the way in which climatic factors influence colonization of microorganisms and their clearance from the respiratory tract. Factors influencing pathogen survival in the external environment should also be studied in more detail. In this connection, the biological consequences of widely fluctuating climatic conditions should also be emphasized.

Data discussed in this chapter clearly indicate that disease under natural conditions is a result of complex interactions between several factors related to the host and the pathogen. Multifactorial experiments are therefore necessary to determine the relative importance of each of the individual factors. Few such studies have yet been performed.

One example of interactions that should be considered is the relationship between air temperature and relative humidity. Several studies have demonstrated the effects of RH on pathogen survival and local defence of the upper respiratory tract. More research should be done to study the effects of RH at different environmental temperatures.

Meteorological forecasting is an important approach in the efforts to prevent health problems associated with weather conditions. The aim of forecasting animal disease should be to provide a service to farmers and veterinarians and aid them in making strategic and tactical decisions related to housing and management. Meteorological forecasting of animal health and disease has been discussed by Hugh-Jones and Yvore (1989).

Models for meteorological forecasting of parasitic conditions have been applied (Smith 1970). Prediction models to forecast and analyze the airborne spread of foot and mouth disease have also been developed (Donaldson et al. 1982).

Little progress has yet been made in the compilation of existing data on the relationships between adverse thermal environment and animal health with regard to forecasting adverse weather conditions in order to reduce the impact of multifactorial microbial diseases as discussed in this chapter. However, field observations made by farmers and veterinarians and data from epidemiological studies (Sect. 2) clearly indicate that such an approach is relevant. This is particularly the case for respiratory disease, in several animal species, which has been associated with cold, in particular the sudden change in weather to cool and humid conditions during the winter season. Models for prediction of such changes in weather will enable farmers to employ relevant management strategies in order to minimize the consequences of adverse weather conditions.

Although knowledge is still inadequate with respect to thermal requirements associated with health criteria, efforts should be made to develop models for rational strategic and tactical decisions to forecast and prevent disease. Because of insufficient data, such models should initially be based on epidemiological and applied research data and practical experience, and later be refined as knowledge from basic research becomes available.

References

Alberts JO, Graham R (1948) Fowl cholera in turkeys. N Am Vet 29: 24–26

Anderson IB, Lundquist GR, Proctor DF (1972) Human nasal mucosal function under four controlled humidities. Am Rev Resp Dis 106: 438–449

Andresen U, Horsten D, Wiecha B (1981) Der Einfluss Grosswetterlage auf das Entstehen der "crowding disease" und enzootische Bronchopneumonie der Kälber sowie ihre Bekämpfung mit Imuresp-p. Dtsch Tierärtl Wochenschr 88: 107–112

Andrews CH (1964) The complex epidemiology of respiratory virus infections. Science 146: 1274–1277

Armstrong WD, Cline TR (1977) Effects of various nutrient levels and environmental temperatures on the incidence of colibacillary diarrhoea in pigs: intestinal fistulation and titration studies. J Anim Sci 45: 1042–1049

Baetjer AM, Lowry RH, Bang FB (1960) Effects of environmental temperature and humidity on spread of virus in the respiratory tract. Fed Proc 19: 178

Ben Nathan D, Heller ED, Perek M (1976) The effect of short heat stress upon leucocyte count, plasma corticosterone level, plasma and leucocyte ascorbic acid content. Br Poult Sci 17: 481–485

Blalock JE (1989) A molecular basis for bidirectional communication between the immune and neuroendocrine systems. Phys Rev 69: 1–32

Blecha F, Kelley KW (1981a) Cold stress reduces the acquisition of colostral immunoglobulin in piglets. J Anim Sci 52: 594–600

Blecha F, Kelley KW (1981b) Effects of cold and weaning stressors on the antibody-mediated immune responses of pigs. J Anim Sci 53: 439–447

Blood DC, Radostits OM, Henderson JA (1983) Veterinary medicine, 6th edn. Balliere Tindall, London, p 755

Brandly CA, Moses HE, Jones EF, Jungherr EL (1946) Epizootiology of Newcastle disease of poultry. Am J Vet Res 7: 243–249

Breazile JE (1988) The physiology of stress and its relationship to mechanisms of disease and therapeutics. Vet Clin North Am Food Anim Pract 4: 441–480

Brown RW, Thomas JL, Cook HM, Riley JL, Booth GD (1977) Effect of environmental temperature stress on intramammary infections of dairy cows and monitoring of body and intramammary temperatures by radiotelemetry. Am J Vet Res 38: 181–187

Cheville NE (1979) Environmental factors affecting the immune response of birds – a review. Avian Dis 23: 308–317
Claman HN (1987) Corticosteroids – immunologic and anti-inflammatory effects. In: Berczi I, Kovacs K (eds) Hormones and immunity. MTH Press, Lancaster, pp 38–42
Curtis SE, Drummond JG (1982) Relative and quantitative aspects of aerial bacteria in animal houses. In: Recheigel M (ed) Handbook of agricultural productivity, vol II. CRC Press, Cleveland, pp 107–115
Curtis SE, Morris GL (1982) Operant supplemental heat in swine nurseries. Proc 2nd Int Livestock Symp American Society of Agricultural Engineers, St Joseph, Michigan, pp 295–297
Curtis SE, Kingdon DA, Simon J, Drummond JG (1976) Effects of age and cold on pulmonary bacterial clearance in the young pig. Am J Vet Res 37: 299–301
Dennis MJ (1986) The effects of temperature and humidity on some diseases – a review. Br Vet J 142: 472–485
Dividich J le, Noblet J le, Aumaitre A (1982) Environmental requirements of early-weaned intensively reared piglets. Proc 2nd Int Livestock Symp American Society of Agricultural Engineers, St Joseph, Michigan, pp 353–361
Donaldson AI (1978) Factors influencing the dispersal, survival and deposition of airborne pathogens of farm animals. Vet Bull 48: 83–94
Donaldson AI, Ferris NP (1976) The survival of some airborne viruses in open air conditions. Vet Microbiol 1: 413–420
Donaldson AI, Gloster J, Harvey LDJ, Deans DH (1982) Use of prediction models to forecast and analyse airborne spread during the foot-and-mouth disease outbreaks in Brittany, Jersey and the Isle of Wight in 1981. Vet Rec 110: 53–57
Dougherty (1971) Environment and physiopathology. In: A guide to environmental research on animals. Committee on physiological effects of environmental factors on animals. Agricultural Board, National Research Council, National Academy of Sciences, Washington DC, p 226
Elzhary M, Derbyshire JB (1979) Effect of temperature, relative humidity and medium on the aerosol stability of infectious bovine rhinotracheitis virus. Can J Comp Med 43: 158–167
Ewbank R (1968) An experimental demonstration of the effect of surface cooling upon the health of the bovine mammary gland. Vet Rec 83: 686–688
Ferguson W, Branagan D (1972) Meteorological effects on animal disease. In: Tromp SW (ed) Progress in biometeorology, division A, progress in human biometeorology. Swets & Zeitlinger, Amsterdam, pp 93–107
Fetrow JF, Madison JB, Galligan D (1985) Economic decision in veterinary practice: a method for field use. J Am Vet Med Assoc 186: 792–797
Furuuchi S, Shimizu Y (1976) Effect of ambient temperatures on multiplication of attenuated transmissible gastroenteritis virus in the bodies of newborn piglets. Infect Immun 13: 990–992
Germain R, Redon P, Tournut J (1975) Role des facteurs climatique dans l'etologie des infections a myxovirus parainfluenza III dans la region Midi-Pyrinees. Rev Med Vet 126: 329–339
Gloster J, Seller RF, Donaldson AI (1982) Long distance transport of foot-and-mouth disease over the sea. Vet Rec 110: 47–52
Griffin JF (1989) Stress and immunity: a unifying concept. Vet Immunol Immunopathol 20: 263–312
Gross WB, Siegel HS (1988) Environment-genetic influences on immunocompetence. J Anim Sci 66: 2091–2094
Hahn GL (1981) Housing and management to reduce climatic impacts on livestock. J Anim Sci 52: 175–186
Hahn GL (1985) Management and housing of farm animals in hot environments. In: Yousef MK (ed) Stress physiology in livestock, vol II. Ungulates. CRC Press, Boca Raton, pp 151–174
Harper GJ (1961) Airborne micro-organisms: survival tests with four viruses. J Hyg 59: 479–486
Heller ED, Ben Nathan D, Perek M (1979) Short heat stress as an immunostimulant in chicks. Avian Pathol 8: 195–203
Holmes CW (1971) Local cooling of the mammary gland and milk production in the cow. J Dairy Res 38: 3–7
Hropot M (1970) Untersuchungen über den Einfluss des Wetters auf die Entstehung der akuten Mastitis des Rindes. Vet Diss, Ludwig-Maximilians-Universität, München

Hudson DB (1931) The influence of environmental temperature on the mortality in chicks inoculated with the virus of infectious bronchitis. Poult Sci 10: 391

Hugh-Jones ME (1976) Effects of heat and cold on diseases in sheep. In: Tromp SW, Progress in biometeorology, division B, progress in animal biometeorology. Swets & Zeitlinger, Amsterdam, pp 480–486

Hugh-Jones ME (ed) (1989) Animal health and production at extremes of weather. World Meteorological Organization. Technical note 191. Secretariat of the World Meteorological Organization, Geneva

Hugh-Jones ME, Yvore P (1989) Some aspects of the meteorological forecasting of animal health and diseases. In: Hugh-Jones ME (ed) Animal health and production at extremes of weather. World Meteorological Organization. Technical note 191. Secretariat of the World Meteorological Organization, Geneva

Jericho KWF, Darcel CQ (1978) Response of the respiratory tract of calves kept at controlled climatic conditions to bovine herpesvirus 1 in aerosol. Can J Comp Med 42: 156–167

Jericho KWF, Magwood SE (1977) Histological features of respiratory epithelium of calves held at differing temperatures and humidity. Can J Comp Med 41: 369–379

Jones CDR (1987) Proliferation of *Pasteurella haemolytica* in the calf respiratory tract after an abrupt change in climate. Res Vet Sci 42: 179–186

Jones CDR, Bull JR (1987) Deposition and clearance of radiolabelled polystyrene spheres in the calf lung. Res Vet Sci 42: 82–91

Jones CDR, Webster AJF (1981) Weather induced changes in airborne bacteria within a calf house. Vet Rec 109: 493–494

Jones CDR, Webster AJF (1984) Relationships between counts of nasopharyngeal bacteria, temperature, humidity and lung lesions in veal calves. Res Vet Sci 37: 132–137

Juszkiewicz T (1967) Experimental *Pasteurella multocida* infection in chickens exposed to cold: biochemical and bacteriological investigations. Pol Arch Weter 10: 615–625

Juszkiewicz T, Cakalowa A, Stefaniakowa B, Madejski Z (1967) Experimental pasteurella infection in normal and chlorpromazine-premedicated cockerels, subjected to heat stress. Pol Arch Weter 10: 601–614

Kelley KW (1980) Stress and immune function: a bibliographic review. Ann Rech Vet 11: 445–478

Kelley KW (1982) Immunobiology of domestic animals as affected by hot and cold weather. Proc 2nd Int Livestock Symp American Society of Agricultural Engineers, St Joseph, Michigan, pp 470–482

Kelley KW (1985) Immunological consequences of changing environmental stimuli. In: Moberg GP (ed) Aimal stress. American Physiological Society, Bethesda, Maryland, pp 193–223

Kelley KW (1988) Cross-talk between the immune and endocrine systems. J Anim Sci 66: 2095–2108

Kelley KW, Blecha F, Regnier JF (1982a) Cold exposure and absorption of colostral immunoglobulins by neonatal pigs. J Anim Sci 55: 363–368

Kelley KW, Greenfield RE, Evermann JF, Parish SM, Perrymann LE (1982b) Delayed-type hypersensitivity, contact sensitivity, and phytohemagglutinin skin-test responses of heat- and cold-stressed calves. Am J Vet Res 43: 775–779

Khansari DN, Murgo AJ, Faith RE (1990) Effects of stress on the immune system. Immunol Today 11: 170–175

Kirton AH, O'Hara GJ, Shortridge EH, Cordes DO (1976) Seasonal incidence of enzootic pneumonia and its effect on the growth of lambs. N Z Vet J 24: 59–64

Knight V, Gilbert BE, Wilson SZ (1985) Airborne transmission of virus infections. Banbury Report 22: Genetically altered viruses and the environment 11, pp 73–93

Kopecky KE, Pugh GW, McDonald TJ (1981) Influence of outdoor winter environment on the course of infectious bovine keratoconjunctivitis. Am J Vet Res 42: 1990–1992

Lechtworth GJ, Carmichael LE (1984) Local tissue temperature: a critical factor in the pathogenesis of bovid herpesvirus 2. Infect Immun 43: 1072–1079

Lehner B, Sasshofer K (1984) Untersuchungen Zur Änderung der Keimflora im Schweinstall in Abhängigkeit zur Stallraumtemperatur unter experimentellen Bedingungen. Dtsch Tierärztl Wochenschr 91: 222–224

MacVean DW, Franzen DK, Keefe TJ, Bennett BW (1986) Airborne particle concentration and meterologic conditions associated with pneumonia incidence in feedlot cattle. Am J Vet Res 47: 2676–2682

Martin SW, Schwabe CW, Franti CE (1975) Dairy calf mortality rate: influence of meteorologic factors on calf mortality rate in Tulare county, California. Am J Vet Res 36: 1105–1109

McIlroy SG, Goodall EA, McCracken RM, Stewart DA (1989) Rain and windchill as factors in the occurrence of pneumonia in sheep. Vet Rec 125: 79–82

Mendoza Sarmiento J (1986) Mechanism of cold-induced increase in susceptibility to enterotoxigenic *Escherichia coli*-induced diarrhoea of the newborn pig. Diss Abst Int B 47(2): 529

Mitchell PJ, Hooper PT, Collyer DN (1981) Heat stress and diarrhoea in neonatal calves. Aust Vet J 57: 392

Moberg GP (1985) Biological response to stress: key to assessment of animal well-being? In: Moberg GP (ed) Aimal stress. American Physiological Society, Bethesda, Maryland, pp 27–49

Nelson FE, Schuh JD, Stott GH (1967) Influence of season on leucocytes in milk. J Dairy Sci 50: 978–979

Olson DP, Bull RC (1986) Antibody responses in protein-energy restricted beef cows and their cold stressed progeny. Can J Vet Res 50: 410–417

Olson DP, Papasian CJ, Ritter RC (1980a) The effect of cold stress on neonatal calves. I. Clinical conditions and pathological lesions. Can J Comp Med 44: 11–18

Olson DP, Papasian CJ, Ritter RC (1980b) The effect of cold stress on neonatal calves. II. Absorption of colostral immunoglobulins. Can J Comp Med 44: 19–23

Olson DP, Bull RC, Woodard LF, Kelley KW (1981) Effects of maternal nutritional restriction and cold stress on young calves: absorption of colostral immunoglobulins. Am J Vet Res 42: 876–879

Olson LD, McCune EL, Bond T (1968) Epizootiologic pattern of fowl cholera in turkeys in Missouri. Proc 12th Annu Meet, US Livestock Sanitary Assoc, pp 239–243

Pouteaux VA, Christison CI, Rhodes CS (1982) The involvement of dietary protein and chilling in the etiology of diarrhoea in newly weaned pigs. Can J Anim Sci 62: 1199–1209

Pririe HM (1981) Acute fatal pneumonia in calves due to respiratory syncytial virus. Vet Rec 109: 87

Pritchard DG, Carpenter CA, Morzaria SP, Harkness JW, Richards MS, Brewer JI (1981) Effect of air filtration on respiratory disease in intensively housed veal calves. Vet Rec 109: 5–9

Raddatz RL, Hammond GW, Gelskey DE (1991) Long range atmospheric transport of aerolized influenza virus. 10th Conf on Biometeorolgy and aerobiology, special session on hydrometeorlogy. American Meteorlogical Society, Boston, pp 37–41

Rafai P, Kovacs F, Ballasch A (1985) Influence of microclimatic factors on the immunoglobulin status of newborn calves kept singly in outdoor hutches. Proc 5th Int Congr Animal hygiene, Hannover, pp 307–312

Rawson RE, Dziuk HE, Good AL, Anderson JF, Bates DW, Ruth GR (1988) Thermal insulation of young calves exposed to cold. Can J Vet Res 53: 275–278

Regnier JA, Kelley KW (1981) Heat- and cold-stress suppresses in vivo and in vitro cellular immune responses of chickens. Am J Vet Res 42: 294–299

Regnier JA, Kelley KW, Gaskins CT (1980) Acute thermal stressors and synthesis of antibodies in chickens. Poult Sci 59: 985–990

Robinson JH, Easterday BC, Tumova B (1979) Influence of environmental stress on avian influenza virus infection. Avian Dis 23: 346–354

Roth JA (1985) Cortisol as a mediator of stress-associated immunosuppression in cattle. In: Moberg GP (ed) Animal stress. American Physiological Society, Bethesda, Maryland, pp 225–243

Roussel JD, Ortego JD, Gholson JH, Frye JB (1969) Effect of thermal stress on the incidence of abnormal milk. J Dairy Sci 52: 912

Sanford SE, Josephson GKA (1981) Porcine *Haemophilus pleuropneumonia* cpizootic in southwestern Ontario: clinical, microbiological, pathological and some epidemiological findings. Can J Comp Med 45: 2–7

Schalm OW, Caroll EJ, Jain NC (1971) Bovine mastitis. Lea & Febiger, Philadelphia, p 38

Scheepens CJM (1991) Effects of draught as climatic stressor on the health status of weaned pigs. Thesis, University of Utrecht

Shimizu M, Shimizu Y, Kodama Y (1978) Effects of ambient temperatures on induction of transmissible gastroenteritis in feeder pigs. Infect Immun 21: 747–752

Shope RE (1955) The swine lungworm as a reservoir and intermediate host for swine influenza virus.

V. Provocation of swine influenza by exposure of prepared swine to adverse weather. J Exp Med 102: 567–572

Siegel HS (1974) Environmental stress and animal health: a discussion of the influence of environmental factors on the health of livestock and poultry. Proc 14th Int Livestock Environ Symp, American Society of Agricultural Engineers, St Joseph, Michigan, pp 14–20

Simensen E (1974) The relationship between weather and incidence of paturient paresis and mastitis in dairy cows. Nord Vet Med 26: 382–386

Simensen E (1985) Diseases of poultry in hot and cold environments. In: Yousef MK (ed) Stress physiology in livestock. CRC Press, Boca Raton, pp 137–146

Simensen E, Norheim K (1983) An epidemiological study of calf health and performance in Norwegian dairy herds. III. Morbidity and performance: literature review, characteristics. Acta Agric Scand 33: 1–8

Simensen E, Olson LD, Hahn GL (1980) Effects of high and low environmental temperatures on clinical course of fowl cholera in turkeys. Avian Dis 24: 816–832

Sinha SK, Hanson RP, Brandly CA (1957) Effect of environmental temperature upon facility of aerosol transmission of infection and severity of Newcastle disease among chickens. J Infect Dis 100: 162–168

Sinovic B (1970) The influence of environmental temperature, age, sex and breed on the mortality from infectious laryngotracheitis. 14th World Poultry Congr 3: 285–290

Slavik MF, Skeeles JK, Beasley JN, Harris GC, Roblee P, Hellwig D (1981) Effect of humidity on infection of turkeys with *Alcaligenes faecalis*. Avian Dis 25: 936–942

Smith LP (1970) Weather and animal diseases. World Meteorological Organization, Technical note 113, Secretariat of the World Meteorological Organization, Geneva

Songer JR (1967) Influence of relative humidity on the survival of some airborne viruses. Appl Microbiol 15: 35–42

Stott GH, Wiersma F, Menefee BE, Radwandski FR (1976) Influence of environment on passive immunity in calves. J Dairy Sci 59: 1306–1311

Subba Rao DS, Glick B (1970) Immunosuppressive action of heat in chickens. Proc Soc Exp Biol Med 133: 445–448

Thaxton P, Siegel HS (1970) Immunodepression in young chickens by high environmental temperature. Poult Sci 49: 202–205

Thomson RG, Glinka F (1974) A brief review of pulmonary clearance of bacterial aerosols emphasizing aspects of particular relevance to veterinary medicine. Can Vet J 15: 99–104

Thrusfield M (1988) The application of epidemiological techniques to contemporary veterinary problems. Br Vet J 144: 455–469

Wathes CM (1987) Airborne microorganisms in pig and poultry houses. In: Bruce JM, Sommer M (eds) Environmental aspects of respiratory diseases in intensive pig and poultry houses, including the implications for human health. Commission of the European Communities, Luxembourg, pp 57–71

Wathes CM, Howard K, Webster AJF (1985) Survival of *Escherichia coli* bacteria. Proc 5th Int Congr Animal hygiene, Hannover, pp 175–181

Webster AJF (1981) Weather and infectious disease in cattle. Vet Rec 108: 183–187

Webster AJF (1983) Environmental stress and the physiology, performance and health of ruminants. J Anim Sci 57: 1584–1593

Whittlestone P (1976) Effect of climatic conditions on enzootic pneumonia of pigs. Int J Biometeorol 20: 42–48

Wiseman A, Selman IE, Pririe HM, Harvey IM (1976) An outbreak of acute pneumonia in young, single-suckled calves. Vet Rec 98: 192–195

Witte KH (1986) Schweineinfluenza (Pathogenese, Epidemiologie, Nachweis). Prakt Tierarzt 7: 592–598

Wood RL (1986) Erysipelas. In: Leman AD, Straw B, Glock RD, Mengeling WL, Penny RHC, Scholl E (eds) Diseases of swine. Iowa State University Press, Ames, pp 571–583

Yoder HW, Drury LN, Hopkins SR (1977) Influence of environment on airsacculitis: effects of relative humidity and air temperature on broilers infected with *Mycoplasma synoviae* and infectious bronchitis. Avian Dis 21: 195–202

3 Thermoelectric Methods for Measurement of Sap Flow in Plants

Y. Cohen

1 Introduction

Thermoelectric methods for the determination of sap flow in stems use a heat pulse, given to stems from external sources, as a tracer for water movement in the plant. The methods were developed to improve the assessment of transpiration from whole plants both for the study of plant water relationship and to improve irrigation management. With the increasing shortage of water in many regions around the world, the necessity for accurate knowledge on water budget and plant–water requirements has became very important. In crop modeling, the input of accurate data on water use, based on relatively short time intervals, is necessary for an improved understanding of the environmental effects on crop production.

Numerous attempts have been made to develop methods for the accurate estimation of evapotranspiration. The majority of the methods are indirect, estimating the soil water balance or water vapor flux above the canopy. Micrometeorological methods remain the best approach for estimating transpiration in field crops, but in orchards the heterogeneities introduced by shelter belts and spaced planting make such methods difficult to apply. Most direct methods of measurement (lysimetery, tent enclosures, cuvette or porometry) cause some interference with the plant and require complex equipment, while some are difficult to extrapolate for a whole plant or an entire canopy. These methods are, therefore, generally unsatisfactory for the study of the plant–water relationship, irrigation management and accurate water budget. The idea of direct measurement of transpiration in plants is attractive because it overcomes some of the basic problems of indirect measurement. For example, the technique is principally independent of environmental conditions, canopy structure, and root system characterization. The method has the advantage of being representative of the whole plant with a simple, yet theoretically sound basis. The transpiration from a single plant can be integrated over the entire canopy. The measurement can be used under field conditions and may be repeated for the same plant, generally over a period of days or weeks, thereby allowing for an evaluation of temporal transpiration changes without disturbing the natural development of the plant. Recent developments in instrumentation, related to this method, have made the automatic long-term measurement of transpiration in the field feasible.

The thermoelectric method for determining sap flow has been studied for nearly 60 years but intensive investigations by many workers and significant progress in the technology of the method have been made particularly in the past 15 years. Numerous attempts to improve the theoretical understanding of the method have been made together with the enhanced use of technological innovations; however, although the method has spread to a large number of laboratories during the last 15 years, if has not yet been established for large-scale use in irrigation management for two reasons: (1) The application of the method in different species requires either a calibration or test for verification of the theory that is not available for many species. (2) The systems built by individual laboratories, as well as those presently commercialized, are highly sensitive, requiring delicate handling for trouble-free operation.

Over the years of investigation, a few models were developed for the solution of heat dissipation equations quantifying sap flow in the xylem tissue. Subsequently, several approaches and system designs were developed. The general principles of these approaches have been described recently by Dugas (1990b). In the following sections, the physical principles on which the various approaches are based will be described, and the applicability of the different techniques will be evaluated. Comprehensive studies made by many workers on the verification of the theoretical models by experimental results for different species will be discussed. Finally, results related to the application of the method in field studies to investigate the plant's response to its environment and the potential use of the method for improving farm management will be reviewed.

2 Theory

Two different approaches, based on the solution of the convective heat flow equation in the stem with water-conducting elements, have been used for sap flow measurement. By one approach, sap flow is calculated from heat velocity measurements, estimated from the time required for the heat pulse to move from a heater inserted in the stem to a downstream location. This method is known as the "heat pulse method". The second approach calculates the mass flow of water as a residual in the heat balance of a stem with or without the insertion of the sensors into the stem. This method is known as the "steady-state heat balance method". Several variations in the physical principles or in the system measurement design of both methods have been applied. In this section the concepts and modifications of the methods will be discussed.

2.1 Heat Pulse

The method of measuring heat velocity in plants was first applied by Huber and his colleagues (Huber 1932; Huber and Schmidt 1937). Their approach equated the heat velocity through the stem to the sap velocity. Marshall (1958), providing a firm physical basis for the relationship between heat and sap velocity, showed that the concept used by earlier research was incorrect. Most of the original studies on the heat pulse method for estimating sap velocity were conducted for large trees. The large size of the tree trunks required the insertion of the temperature sensors into the trunk to detect temperature changes, which were not accurately obtained by surface temperature measurement.

Heat velocity in the stem was correlated with the rate of water loss or environmental variables. The important contribution of those earlier works was comprehensively summarized by Leyton (1970), however, quantification of sap flow has not been made due to either an incorrect application of the theory for heat dissipation in the stem or the lack of advanced technology.

The model proposed by Marshall (1958) considers the tree as a semi-infinite half-space. It solves the case of heat pulse delivered by a line source inserted radially into the tree, and assumes a two-dimensional heat flow. In this model, the temperature elevation T (K), produced by the heat pulse after time t (s), and at a distance x (mm) downstream from the line heater, is given by:

$$T = [Q/4\pi\rho ckt)] \exp[-(x - vt)^2/(4kt)], \quad (1)$$

where Q is the heat output per unit length of the heater ($J\ mm^{-1}$), ρ ($mg\ mm^{-3}$), c ($J\ mg^{-1}\ K^{-1}$), and k ($mm^2\ s^{-1}$) are the density, specific heat, and thermal diffusivity of the wood, respectively; and v is the convective heat velocity ($mm\ s^{-1}$). Placing the line heater and the temperature probe in the same diametrical, longitudinal plane simplifies the solution to an apparent one-dimensional form as given in Eq. (1). The function defined by Eq. (1) has a maximum occurring at time t_m when the first derivative of the equation equals zero. This condition yields:

$$v = (x^2 - 4kt_m)^{0.5}/t_m. \quad (2)$$

The measurement of t_m in Eq. (2) allows the computation of the sap velocity J_I ($mm\ s^{-1}$) by:

$$v = (\rho_I c_I/\rho c) J_I, \quad (3)$$

where ρ_I and c_I are the density and specific heat of the sap. Marshall (1958) used the temperature-time curve of a single heat pulse for computation of k, but when $v = 0$, Eq. (2) yields:

$$k = x^2/4t_m. \quad (4)$$

Consequently, k can be determined when no convective transport is taking place, which is likely to be at predawn. The test is done by inserting two thermocouple junctions into the trunk equidistant from a heater. The differential output from the thermocouple following a heat pulse should equal zero (Cohen et al. 1981).

In order to determine the volumetric flow, F ($mg\ s^{-1}$), the sap velocity (J_I) must be integrated over the cross-sectional area of the stem:

$$F = \int J_I d_s, \quad (5)$$

where d_s is the element of stem area in which J_I has been determined.

Closs (1958) analyzed the heat transport in the cotton stem. Although his solution is based on Marshall's model, he proposed to solve the equation for one-dimensional flow. His solution involved using a differential temperature measurement at two asymmetrically located points above and below the heat source. In this configuration, the heat velocity is given as:

$$v = (x_1 - x_2)/2t_0, \quad (6)$$

in which x_1 an x_2 are the distances above and below the line heat source, respectively, and t_0 is the time required for the temperature difference between x_1 and x_2 to return to its initial value. Closs's solution has been widely applied in trees by many researchers, as knowledge of k is not required (Ladefoged 1960; Morikawa 1972; Swanson 1972; Lassoie et al. 1977). However, a serious error in the measurement may be introduced by this approach if the sensors above and below the heater are not located accurately in the stem and if their time constant is not identical. Marshall's solution considered sapwood in the stem to be comparatively homogeneous. But in most conifers and ring-porous hardwoods, the sap moves up a number of annual rings at a different velocity (Kozlowski et al. 1965); therefore, Marshall's theoretical model is inadequate, leaving unknown factors in the solution (Pickard 1973). In addition, the heater and the temperature sensor inserted into the trunk have thermal properties differing from those of the surrounding wood, and they interrupt the sap flow in their immediate neighborhood. Therefore, when the equation for heat transport is based on Marshall's idealized heat transport model, it seriously underestimates actual sap velocity in the stem (Marshall 1958; Doley and Grieve 1966; Leyton 1970).

Swanson and Whitfield (1981) also showed the deviation of measured sap velocity from the idealized theory. They developed a numerical solution of the flow of heat and sap through sapwood to correct the data. The correction includes the blockage caused by implanted probes, the wound width, the probe spacing, and the thermal properties of the sensors. Their computer simulation of the fluxes consists of fitted coefficients derived for particular probes. Swanson (1983) carried out several experiments that verified the numerical model. The results of later studies using the proposed correction provided support for this technique (Edwards and Warwick 1984; Green and Clothier 1988; Hatton et al. 1990); however, heterogeneity among species, whether natural or caused by implantation of the sensors in the conducting tissues, increases uncertainties about the coefficients used for the correction of v (Green and Clothier 1988).

Cohen et al. (1981) proposed the use of an experimental calibration capable of being carried out either under field conditions or in the laboratory. The calibration, using a weighing method, yielded a coefficient incorporating several variables: the structure of the water-conducting system, a possible error in estimating the active xylem area, effect of the wound, thermal quality of the sensor, and configuration of the physical system for sap velocity measurement.

The ultimate goal of the measurement of sap velocity is the sap flow rate; therefore, the area of active xylem should be determined in order to evaluate the sap flux density. The simplifying assumption by earlier workers that sap velocity is constant across the sapwood conducting area does not hold for most of the trees (Mark and Crews 1973; Lassoie et al. 1977; Miller et al. 1980; Cohen et al. 1981, 1985; Hatton et al. 1990). Sampling at several depths into the sapwood is usually necessary to characterize the sap velocity profile.

One approach for an accurate estimation of the sapwood area and the characterization of the radial and orientational variability of the sap velocity in the cross-sectional area is by using a multisensor temperature probe proposed by Cohen et al. (1981). The probe consists of six thermistors, placed 8 mm apart, and capable of measuring at six points simultaneously. A schematic diagram describing the probe and the recording system is shown in Fig. 1. To study the orientation variability in

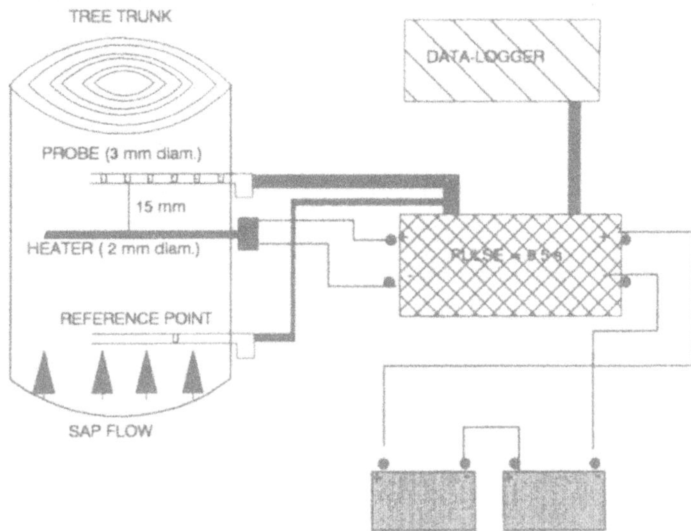

Fig. 1. Configuration of the heat pulse system for sap flow measurement of trees. The probe is made of six thermistors; the reference thermistor is not affected by the heater

the trunk, three probes are inserted at three different orientations of the trunk, making it possible to study 18 points in the cross-sectional area of the trunks.

Another approach for obtaining the integrated sap flux density across the stem is by fitting a least-squares polynomial to the point estimates of sap flow velocity, and then integrating this function across the sapwood area (Edwards and Warwick 1984; Green and Clothier 1988). Hatton et al. (1990) suggested an improved method for integrating sap velocity across the stem section of trees using a weighted average approach. Their approach was shown to be a more robust estimator of flux when velocity profiles exhibit large curvatures.

The physical principles for estimating heat velocity in trees described above were also applied in herbaceous plants, but because of the difference in stem diameter between the two groups, the design of the measurement system and the computation of sap velocity from heat velocity differ. In most of the reported applications of the heat pulse in herbaceous plants, the differential temperature measurement system was used. In order to eliminate the interruption of the stem tissue, the temperature sensors were located on the stem surface above and below the heater (Bloodworth et al. 1955; Closs 1958; Wendt et al. 1965). Stone and Shirazi (1975) perceived a substantial effect of change in air temperature on the accuracy of the measurement; therefore, they proposed the use of a temperature-compensated system to eliminate variations through the use of opposing thermistors in a bridge circuit. Insertion of temperature sensors (0.5 mm external diameter) into the stem was proposed by Cohen et al. (1988) in order to minimize the effects of ambient temperature variations on the measurement and to improve detection of temperature changes in the stem cross-sectional area (Fig. 2). An experimental calibration, similar to that used for

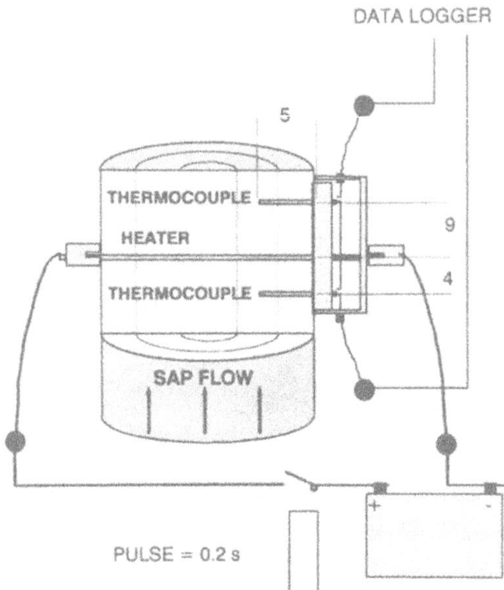

Fig. 2. Configuration of the heat pulse system for sap flow measurement of herbaceous plants

trees, yielded a calibration coefficient with which sap flow is computed, using measured heat velocity and whole stem area.

In the measurement procedure of herbaceous plants, two data are used, allowing accurate determination of convective heat velocity at very low or very high rates. The first recorded datum is the time elapsing from pulse emission to the first re-occurrence of the initial temperature difference, and the second is the time at which the temperature difference is maximum. At low and moderate convective sap velocities in the stem (0–0.17 mm s^{-1}), the first datum can be measured with precision to estimate sap velocity accurately, using Eq. (6). At velocities higher than 0.22 mm s^{-1} the initial differential temperature change is too small and evolves too rapidly for detection when the temperature returns to its initial value, but the temperature wave has a well-defined maximum that is related to sap velocity using Eq. (2). For convective velocities between 0.17 and 0.22 mm s^{-1}, both data can be used.

2.2 *Heat Balance*

Mathematical analysis of the heat balance in a stem segment was first proposed by Vieweg and Ziegler (1960). Leyton (1970) used two permanently heated units (made in thin brass) inserted vertically one above the other in the sapwood of a segment of poplar branch. He obtained a good linear response in the temperature difference between the two units and water flow forced through the segment at different rates.

Based on a similar idea of heat exchange between the sap and a temperature sensor, Granier (1985) used an apparatus consisting of two aluminium probes of 2 mm diameter, which are inserted about 20 mm into the sapwood of the trunk one above the other 100 or 150 mm apart. The upper probe is heated at a constant power by a heating element, while the lower (or reference) probe is at trunk temperature. Each probe contains a copper-constantan thermocouple positioned half-way along its length and both are connected for differential temperature measurement. A calibration set made in several species yielded an equation describing the relationship between temperature differences in the trunk and the heat velocity v (mm s^{-1}):

$$v = 116 \times 10^{-3} (\Delta T_M - T)/\Delta T, \quad (7)$$

in which ΔT and ΔT_M are the temperature differences between the two thermocouples at sap velocity greater than zero or equal zero, respectively.

Sap flow F (mg s^{-1}) is computed by:

$$F = v\, S_A, \quad (8)$$

where S_A is the cross-sectional area of the sapwood (mm^2).

ΔT_M, which is affected by the thermal characteristics of the wood surrounding the heated probe, is estimated over 10-day period by a linear regression of the daily maximum values of ΔT.

The cross-sectional area of the sapwood is estimated by sampling cores from the tree trunk. The sapwood area is distinguished from the heartwood using the contrasting transparency between the two areas in diffuse light. In a pine stand (*Pinus pinaster* cv. Ait.), 33 trees were selected for sampling of the sapwood (Diawara et al. 1991). A linear regression between total cross-sectional area of the sampled trees and sapwood area was fitted on the 33 trees. The total stand cross-sectional area of the sapwood (A_s) was then estimated by:

$$A_s = N[a_s + b(s_t - s_m)], \quad (9)$$

where N is the number of trees per hectare; a_s is the mean cross-sectional area of sapwood per tree (m^2); b is the regression coefficient between the cross-sectional area of sapwood and the total cross-sectional area; s_t and s_m are the mean total cross-sectional area per tree for the whole stand and for the sample (m^2), respectively.

The first successful estimate of mass flow of sap in the trunk of a 50-year-old tree by a complete heat balance was made by Daum (1967). Heat was supplied from four thin heat flux plates inserted underneath the bark at four locations around the trunk. The amount of heat conducted radially to the atmosphere and to the hardwood in the center part of the trunk was monitored. The vertical conduction above and below the heating plates was measured by a temperature gradient between two points. The technique provided a very accurate measurement of mass flow by comparing the mass flow in the trunk to total mass flow in the main branches; however, as in the previous technique described above, an accurate definition of the sapwood area is required. In addition, the insertion of heat flux plates exactly at the borderline between the sapwood and the bark on one side and the hardwood on the other is likely to be impractical.

A modification in the physical basis and a simpler measurement system for estimating total mass flow in trees were proposed later by Cermak et al. (1973, 1982, 1984) and Kucera et al. (1977). Their method (null-balance method) estimates the mass flow of sap by applying a known amount of heat to a stem segment. The convective heat fluxes from the stem sector are measured. This method provides the direct measurement of sap flow, requires neither calibration nor knowledge of the sapwood area or distribution of sap conducting elements in the stem.

The method works in the following way: The stem is heated by five stainless steel electrodes inserted into the stem, and four pairs of thermocouples measure the sap temperature between the electrodes at various depths, using the stem temperature below the electrodes as a reference. The temperature signal from the thermocouple is used to control the heat input from the electrodes in order to maintain a constant temperature difference of 1 K between the heated and nonheated segments.

The mass flow F ($mg\ s^{-1}$) is calculated as:

$$F = [lP/\Delta T\ cd\ (n - 1)] - F_0, \tag{10}$$

where l is the stem circumference (mm); P is the heating power ($J\ s^{-1}$); ΔT is the temperature difference between thermocouples (K); c is specific heat ($J\ mg^{-1}\ k^{-1}$); n and d are the number of the electrodes and the distance between them (mm), respectively; F_0 is assumed constant for similar weather periods because the temperature difference between heated and nonheated tissue is kept constant. F_0 is calculated from Eq. (10) when sap flow assumes zero (at predawn after rain); its value is also taken as the conductive heat loss. To eliminate the effects of environmental variations on the measurement, the stem segment is heat-insulated by a layer of polyurethane foam, protected by an aluminium cover surrounding the tree trunk and protected from precipitation by a polyethylene sheet secured tightly to the trunk. Insertion of the sensors into the stem allows accurate measurements of the temperature in the stem cross section. It is assumed that the temperature gradients between the heated segment and reference point are close to zero. The system has been applied mainly in trees with a large stem diameter, but has also been used in herbaceous species with a modified heating system (Fichtner and Schulze 1990).

This approach has been modified and applied recently to cucumber (*Cucumis sativus* L. cv. chojitsu-oshiai) by Kitano and Eguchi (1989) and to corn, sunflower (*Helianthus annuus*) and potato (*Solanum tuberosum*) plants (Ishida et al. 1991). The system consists of an external heater hugging the stem. The temperature gradient in the stem is measured by inserting a thermocouple into the stem. Conductive heat losses in radial and axial directions are included in the calculation of sap flow. The application in herbaceous species showed that accounting for conductive heat losses improves the accuracy of the measurements, especially at low flow rates. The radial averaging of temperatures by inserting a thermocouple into the stem was shown to result in a significant improvement in the accuracy of the measurements.

Another improved technique for the determination of mass flow by steady-state heat balance is the constant heating power method applied by Sakuratani (1981, 1984) and Baker and van Bavel (1987). This method is based on the application of a known amount of heat to the stem segment from a thin flexible heater that encircles the stem. A schematic diagram of the gauge is shown in Fig. 3.

In this method, conductive heat fluxes up and down the stem are estimated by applying Fourier's law, using the one-dimensional form. The required temperature gradients are estimated from the output of a pair of thermocouples mounted on a cork and placed at the outer surface of the stem above and below the heater. In order to reduce the number of outputs and increase the accuracy of the thermocouple measurement, a differentially wired pair was formed from one junction of each pair

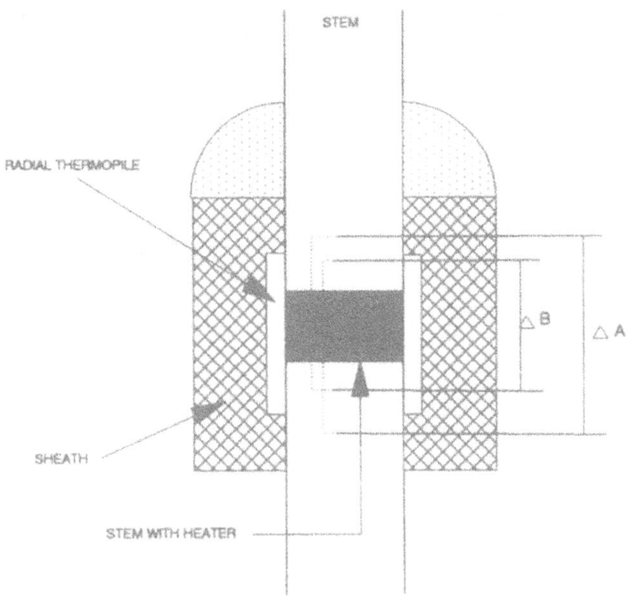

Fig. 3. Configuration of the flow gauge system for the heat balance method (constant heating power). ΔA and ΔB are the temperature differences measured by differential thermocouples A and B, respectively. (Groot and King 1992)

(Steinberg et al. 1990a). Conductive heat fluxes outward through the gauge insulation are estimated by applying the integrated equation for radial flow in a semi-infinite cylinder. The radical temperature gradient is obtained from a thermopile consisting of several thermojunctions in series on either side of a thin sheath surrounding the heater. The difference between total heat input and total conductive fluxes is used for the computation of the mass flow rate of sap F (mg s^{-1}) by:

$$F = [P - LA(\Delta T_u/\Delta_x + \Delta T_d/\Delta_x) - k_g \Delta T_r]/[C_s(T_o - T_i)], \quad (11)$$

in which P is the power to heater (W); L is the stem thermal conductivity (W mm^{-1} K^{-1}); A is the stem area (mm^2); ΔT_u and ΔT_d are the vertical temperature gradients (K) above and below the heater, respectively; Δ_x is the distance (mm) between the two junctions above and below the heater; K_g is the sheath gauge conductance or gauge factor (W k^{-1}); ΔT_r is the temperature difference across the thermopile (K); C_s is the heat capacity of xylem sap (J g^{-1} K^{-1}); T_o is the temperature of sap leaving the heated segment (K); and T_i is the temperature of sap entering the heated segment (K). K_g, representing a "zero set" for each gauge-installation combination, can be calculated when $F = 0$ (Steinberg et al. 1989) or by measuring a severed stem. Dugas (1990a) showed the consistency of K_g calculated from in situ field measurements throughout a growing season for cotton.

In this approach temperature measurement is taken at the outer stem surface, assuming that the radical temperature changes through the stem are negligible. This assumption may limit the range of stem diameters in which the technique may be applied. Representative stem temperature measurements require good contact between gauge and stem. When used for small trees or branches, the surface is treated with sandpaper to remove any loose or rough bark. Lubricant is applied sparingly to the thermojunctions of each gauge to improve contact (Steinberg et al. 1990b).

In the above section two methods, which differ in their physical principles, for estimating sap flow in the stems were described: the heat pulse and the heat balance approaches. The heat pulse method can be applied to trees with large trunks and to herbaceous plants, with a minimum stem diameter of 6 mm using two types of probes. The conversion of measured heat velocity to sap flow requires a calibration that is species-dependent. In some species calibration was eliminated by applying the numerical solution, converting the heat velocity to sap flow. Determination of sapwood area is necessary and several solutions to this problem were proposed.

Accuracy of the measurement is not limited by high flow rates, but at sap velocities lower than 1.4×10^{-2} mm s^{-1} the measurement is not reliable. As the accuracy is dependent on a good contact between the sensor and the tissue, difficulties in the measurement may be found in nonrigid stems. The heater and temperature sensor are inserted into the stem, and no insulation of the probe is required except shading of the stem near the measurement point so as to eliminate heating of the stem by direct radiation. The method has relatively simple electronics, it is portable, self-powered, and fully automated.

The heat balance method does not require calibration for calculation of the sap flow, and knowledge of the active xylem area in the stem is not necessary. It can be applied to trees (if the temperature sensor is inserted into the stem) and to herbaceous species with a stem diameter above a minimum of 5 mm. The method has a long time constant at low flow rates, which is relevant for short-term measurements. The electronic equipment for the measurement is relatively simple, portable and fully automated. The constant heating power method has been used for herbaceous species and also for small horticultural species. It has a limited range of stem diameters appropriate for a specific gauge size. Representative stem temperature measurements require good gauge-stem contact and diurnal change in stem diameter may cause poor contact. Under fluctuating environmental conditions and when the sap flow rate is very high or very low, the measurement may not be reliable.

The variable heating power, with constant temperature difference in the measured stem section (null balance), minimizes the error in temperature measurement when high sap flow causes a low temperature signal. This system design also eliminates possible damage to the tissue caused by excess heating under conditions of low flow rates. As this method was applied mostly to trees, with the temperature sensor inserted into the trunk, the stem temperature difference is accurately determined, and the environmental effect on temperature readings is minimized.

3 Calibration or Verification of the Methods

In this section, only those studies attempting to quantify the sap flow rate have been considered. Specific problems related to the application of the methods in different species will be reviewed. Unfortunately, these problems have been largely neglected in many publications, possibly due to the lack of reasonable explanations.

Calibration or verification of the heat pulse and heat balance method has been performed widely for trees and herbaceous species. Verification is necessary as the methods are based on physical models applied in a biological system with dynamic changes, with considerable sensitivity to environmental variables. The major obstacle to verification under natural conditions is the lack of accurate references to which the technique may be compared. Lysimeters have been used in some studies, but the number of lysimeters (especially for trees) is limited. Many results collected in the field were compared with micrometeorological models, ignoring errors introduced by the inability to take the canopy and other resistances involved in the models into consideration. In addition, the extrapolation from the sap flow rate of a single plant to the transpiration rate of an entire canopy is unsatisfactory due to the variations among plants. This variation is species- and environment-dependent, and in some cases differences of up to 100% in sap flow among plants were detected (Granier 1987; Dugas 1990a; Cohen 1991). Most of the tests, especially with herbaceous species, have been made with potted plants located either in a controlled environment or chambers.

3.1 Heat Pulse

Several tests in different species were carried out to verify the numerical solution proposed by Swanson and Whitfield (1981) for converting heat pulse velocity to sap flow. Figure 4 demonstrates the relationship between transpiration rate from pine tree (*Pinus halepensis* Mill.), as measured by a lysimeter, and corrected sap flow using the numerical solution.

The uncorrected sap flow was seriously underestimated in comparison to lysimeter data, but the corrected flow was very close to the 1:1 line. Swanson (1983) successfully applied the numerical solution to several species and suggested that the coefficients derived from his solution are applicable to different tree species, regardless of the different structure of the water-conducting system in the stem.

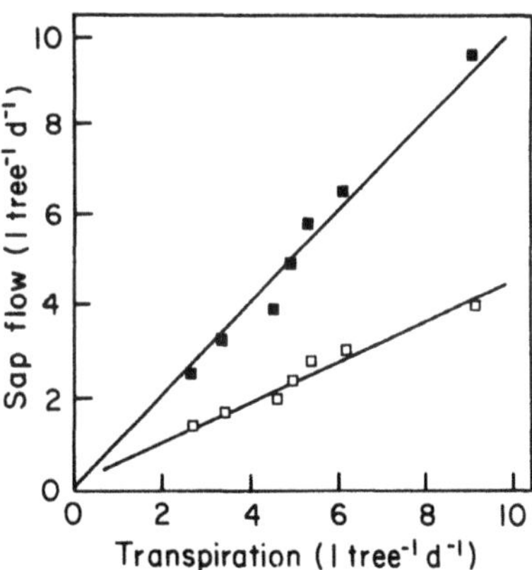

Fig. 4. Transpiration rate of aleppo pine as determined by weighing lysimetry, and sap flow rate calculated from heat pulse velocity. *Open squares* Uncorrected data (before and after rewatering); *filled squares* corrected data (before and after rewatering). (Swanson and Whitfield 1981)

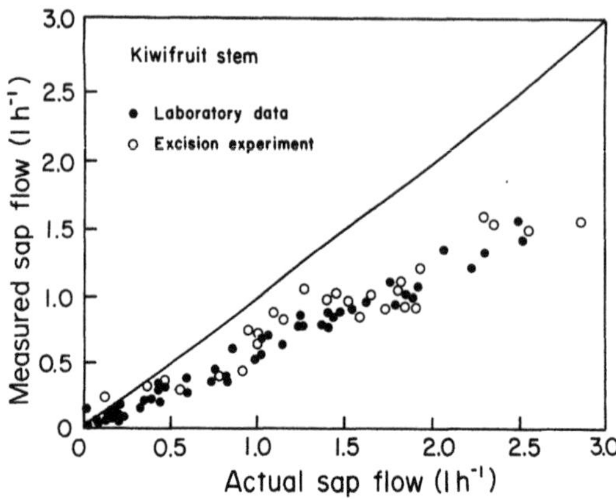

Fig. 5. Laboratory calibration results of three kiwifruit stem sections with a diameter range of 45 to 60 mm. Sap flow was corrected by the numerical solution. Heat pulse measurements underestimate sap flow by a factor of 1.6. (Green and Clothier 1988)

Green and Clothier (1988) used the same solution to correct sap velocities of apple tree (*Malus sylvestris* × Red Delicious) and found very good agreement between actual and calculated sap fluxes. But in kiwifruit vines (*Actinidia deliciosa*), they showed that the corrected sap flow rates, obtained by applying the numerical solution, were lower than the actual rates. The comparison between actual and measured sap flow rates of three kiwifruit stem sections in the laboratory and of an excised kiwifruit vine in the field is shown in Fig. 5. The measured sap flow of both the stem sections and the excised plant was lower by a factor of 1.6 when compared to the actual sap flow.

Inapplicability of the numerical solution to kiwifruit suggests that the anatomy of the stem must also be considered, thus the solution may not be applicable for all species. The theory of heat dissipation in the stem requires that heat be dissipated vertically by a uniform front. The researchers of this study suspected that the sapwood in the kiwifruit stem contains layers of nonconvective elements that require appreciable time to reach thermo-equilibrium with their surroundings, thereby rendering the heat front nonhomogeneous.

Experimental calibration with mature citrus trees (*Citrus sinensis* L.) using three weighing lysimeter systems showed an underestimation of the calculated sap velocity by nearly 45% (Cohen et al. 1981). The proportion of underestimation was similar for all three measured trees and, as shown in Table 1, was consistent regardless of climatic conditions or soil water regime.

This underestimation was explained by the interruption of the conducting elements by the hole in the trunk drilled to accomodate the temperature sensors and the heater. The temperature wave measured in the damaged tissues is a combination of superimposed curves represented by Eq. (1), but with values of v ranging from the actual value in the intact conducting tissues, down to zero in the interrupted vessels. The type of sensors used in this test and the wound size in the trunk differed slightly from those employed in Swanson's tests. Nevertheless, the underestimation of about 45% found in citrus by experimental calibration is comparable with values of 35 to 40% found in conifers using the correction by numerical solution (Swanson and Whitfield 1981). Underestimations found in a laboratory calibration using sections of branches of white poplar (*Populus alba* L.), plane tree (*Platanus orintalis* L.) or

Table 1. Computed sap flow and measured transpiration of adult citrus trees (Cohen et al. 1981)

Environmental conditions	Transpiration measured by lysimeter (1 day^{-1})	Computed sap sap flow from heat pulse (1 day^{-1})	Ratio
Wet soil, clear day	52.0	28.4	0.546
Wet soil, cloudy day	45.0	25.9	0.576
Dry soil, clear day	24.0	13.3	0.554
Dry soil, cloudy day	26.0	14.1	0.542
Wet soil, cloudy day	40.0	21.9	0.548
Mean			0.553
Coefficient of variation			2.4%

Douglas-fir (*Pseudotsuga meziesii* Mirb.) was 46, 43, and 47%, respectively (Cohen et al. 1981, 1985).

In spite of the similar values of the coefficients for several species given above, a different calibration coefficient may be obtained for other species if the anatomical structure of the stem is different. Therefore, it is necessary to conduct an experimental calibration for each species which has not been calibrated. In some species such as palm tree (*Phoenix dactilifera* cv. medjhool), application of the method for sap flow measurement was not possible because of the specific structure of the conducting system. A calibration test in the palm tree was done under conditions of extremely low air humidity and high temperature in the southern part of Israel for two seasons. The test indicated that the water conducting tissue adjacent to the temperature sensors in the stem dried out gradually, and that 3 or 4 days after the insertion of the sensors, sap flow near the sensor ceased completely.

The calibration procedure for herbaceous plants used by Cohen et al. (1988) differs in principle from that applied to trees. In herbaceous species, the measured heat velocity in the stem, rather than the sap velocity, is related to water loss. The heat velocity is measured using Eq. (6) and the differential temperature measurements described by Closs (1958). The requirement for the calibration procedure is that the relationship between heat velocity and transpiration must be linear. The slope of the linear regression line characterizes the proportionality factor between heat velocity and water loss. It has the dimension of area, and is a function of stem area.

A test on a series of cotton plants (*Gosypium hirsutum* L. cv. SJ2) of differing stem diameters was performed in order to derive the relationship between stem area and proportionality factor between heat velocity and water loss (Fig. 6). The slope of this relationship is the calibration coefficient indicating the deviation between measured and actual convective heat velocity and apparent effective conductive area of the stem.

Calibration made for different species showed the coefficient to be species-dependent (Cohen et al. 1988, 1990, 1992), but it can be regarded as constant for any sap flow rate, plant size, or environmental variable. An example of variation in the coefficient among plants is given in Table 2 for a set of calibrations in maize (*Zea mays* cv. kakuteru 901).

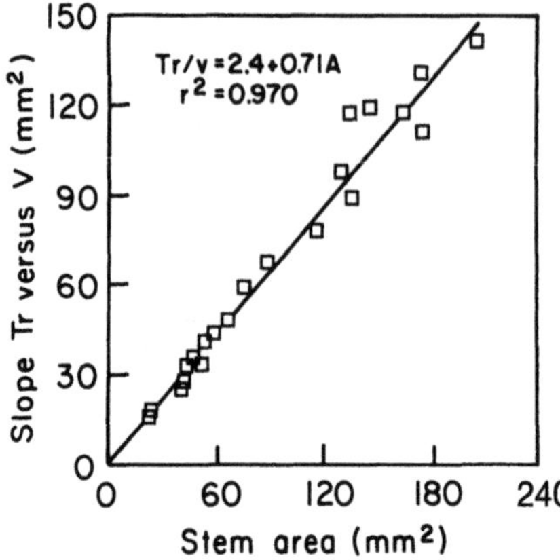

Fig. 6. Relationship between the linear regression slope of transpiration rate (Tr), determined by weight loss vs. convective heat velocity (v) and cross-section area (A) of the stem in 26 cotton plants. (Cohen et al. 1988)

Table 2. Transpiration measured and estimated in corn plants under different environmental and plant conditions (Tr = transpiration, a = stem area, v = heat velocity)

Day of year	Solar radiation ($MJ\,m^{-2}\,day^{-1}$)	Stem area (mm^2)	Tr ($l\,day^{-1}$)	Tr/v × a	Coefficient of correlation R^2
188[a]	11.6	222	0.41	1.64	0.91
189[a]	21.9	234	0.72	1.77	0.87
190[a]	5.4	234	0.22	–	–
191[a]	21.6	234	0.81	1.61	0.93
192[a]	12.4	209	0.41	1.63	0.96
193[a]	11.0	214	0.46	1.67	0.89
195[a]	24.0	203	0.80	1.89	0.93
197[a]	5.6	210	0.31	–	–
198[a]	8.2	210	0.47	1.67	0.89
225[b]	22.5	285	1.21	1.89	0.95
226[b]	22.6	285	1.15	1.82	0.92
227[b]	25.4	282	1.53	1.73	0.91
228[b]	24.4	288	1.27	1.86	0.97
229[b]	22.5	294	1.42	1.65	0.88
230[b]	21.1	288	1.47	1.63	0.91
Average	–	–	–	1.73	–
SD	–	–	–	0.101	–

[a] Seeding on day of year 132.
[b] Seeding on day of year 184.

Despite the difference in plant size and the large variation in radiative flux, differences in the ratio, transpiration (Tr)/[stem area (a) × heat velocity (v)] among plants was small. The dependency of the calibration coefficient on the species indicates that for herbaceous plants, as was shown for trees, the measurement of heat velocity is affected by the distribution of the water-conducting system in the stem. When the method was tested for species having a low stem rigidity (e.g., tomato and squash), the calibration coefficient [Tr/a × v was inconsistent (unpubl. data)]. One possible explanation for the difficulties of measurement in nonrigid stems may be the inadequate contact between the temperature sensor and the tissue (Cohen and Fuchs 1989).

Tests of the heat pulse method show that the method may be applied to trees and herbaceous species with a large range of stem diameters. The minimum tested stem diameter was for a soybean plant (6 mm) and the maximum diameter tested so far was citrus (nearly 300 mm). Generally, one standard-sized sensor can be used for trees or for small stems, without having to adjust a specific sensor to a given stem diameter. The sensor can be used in the trunk of mature trees for a long period (in some species removing it from the trunk may be difficult). In herbaceous species the sensor can be used for the same plant for several weeks if the stem has reached its maximum size. In trees, sapwood with an active xylem area is defined accurately by the six-point probe measurement, as the sap flow across the stem is commonly not uniformly distributed; thus a sampling procedure of cross-sectional area is necessary. Accuracy of the measurement is not affected by the sap flow rate except for very low rates when the heat transport by conductance is of the same magnitude as the convective heat. Accuracy is affected by the voltage and time resolution; full scale voltage is 15 to 20 μV for herbaceous species and 10 to 20 mV for trees. The minimum required voltage resolution is 1×10^{-3} of the full scale; the required time resolution is 3×10^{-1} s. The heater and temperature sensor require no insulation except shading of the stem near the measurement point to eliminate heating of the stem by direct sunlight.

3.2 Heat Balance

Sakuratani (1981) tested the heat balance method with a constant heat power for sunflower (*Helianthus annuus* L. Mammoth Russian) and soybean (*Glycine max* L. Merrill. cv. Miyagishirome) by comparing measured sap flow with transpiration determined by weight loss. The agreement between the two was independent of flow rate up to 120 g h^{-1} and the estimated deviation from the 1:1 ratio was $\pm 10\%$ (Fig. 7).

Similar accuracy of the measurement was reported for potted tomato (*Solanum lycopersimum* cv. Hukuju No. 2) and rice (*Oryza sativa* cv. Nishihomare) (Sakuratani

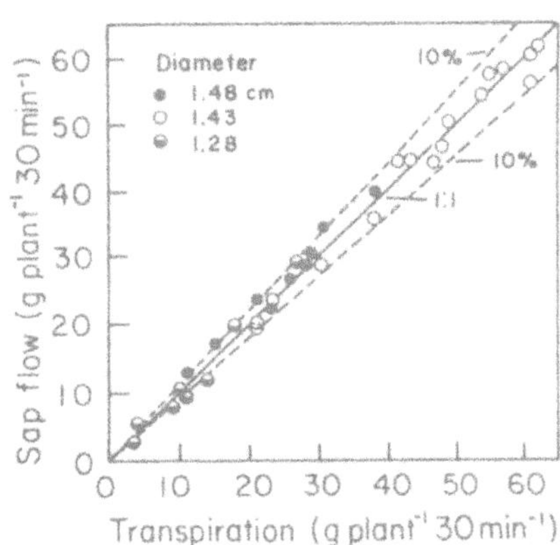

Fig. 7. Transpiration rate of sunflower plants as determined by weight loss, and sap flow rate measured by the heat balance method (constant heating). (Sakuratani 1981)

1984, 1990). Baker and van Bavel (1987) also obtained good agreement between cumulative sap flow and weight loss in potted cotton and sunflower using the same measurement principles. The method was also tested for cotton (cv. GP 3774) potted in the greenhouse and cotton in the field using a 3 m^2 weighing lysimeter with 15 plants m^{-2} (Dugas 1990a). In the greenhouse, the accuracy of the method was comparable with that given above. In the lysimeter, the maximum transpiration rate was about 70 $g\,h^{-1}plant^{-1}$, and the measured flow agreed well with transpiration, but the determination of method accuracy was confounded by high plant-to-plant variability and possibly by errors in measurement at high flow rates. An overestimate of transpiration in sunflower at rates higher than 100 $g\,h^{-1}$ was obtained supposedly because of insufficient length of the heated stem segment (Ham and Heilman 1990).

Steinberg et al. (1989) tested the method for a potted young ficus tree (*Ficus benjamina*) with a 45.2-mm stem diameter in a greenhouse. Their findings showed a very close relationship between cumulative sap flow and weight loss, but when hourly data were compared, a deviation between sap flow and transpiration rate was observed. This deviation was attributed by the researchers to the use of storage water from internal sources in the tree, but it may also be the result of a relatively long time constant of the measurement (Cermak et al. 1984; Baker and van Bavel 1987). A field test of the method using a precision lysimeter was conducted for a 5-year-old, 79-mm stem diameter pecan tree (*Carya illinoensis* cv. "wichita"). The test lasted 12 days with the transpiration rate varying from 100 to 150 $kg\,day^{-1}$. Total sap flow for one of the measured days was 122.8 kg over a 24-h period, corresponding closely to the 113.4 kg of canopy transpiration measured by the lysimeter (Steinberg et al. 1990b). Results of the comparison were not reported for other days, and it is not clear whether this close relationship was maintained over the range of transpiration rates. In the above tests of the heat balance with constant heating power and variable temperature difference, the temperature sensors were placed on the stem surface. This technique assumes that surface temperature is representative of the temperature in the stem cross-sectional area, that the gauge-stem contact is good, and that diurnal changes in stem diameter do not impair contact.

Stem surface temperature may be seriously affected by environmental conditions even though the measured section is well protected. Shakel et al. (1992) had difficulties in their attempt to verify the method for young peach trees. Their analysis indicated that ambient conditions can impose a bias in the gauge signal and, hence, influence gauge accuracy. Groot and King (1992) used a numerical model and sap flow measurements with this method to examine the effect of environmental parameters on the accuracy of the measurement. Flow gauges were constructed and used to measure sap flow rates in seedlings of black spruce (*Picea mariana* cv. Mill.) and jack pine (*Pinus banksiana* cv. Lamb). The model indicated that for flow rates typical of seedlings of these species, shortwave radiation does not cause a significant error in measurement of the across-heater temperature difference (ΔT_r). However, when no reflective covering was used on the gauges, shortwave radiation produced orientation-dependent deviation in modeled radial temperature gradients of up to 50% of the mean value. Gauge accuracy was poor at low flow rates because the convective heat flux and ΔT_r are small, and because the transient response of the gauge is slow. Successful tests were shown by the second approach, the steady-state heat balance method, in which the temperature difference is maintained at a constant

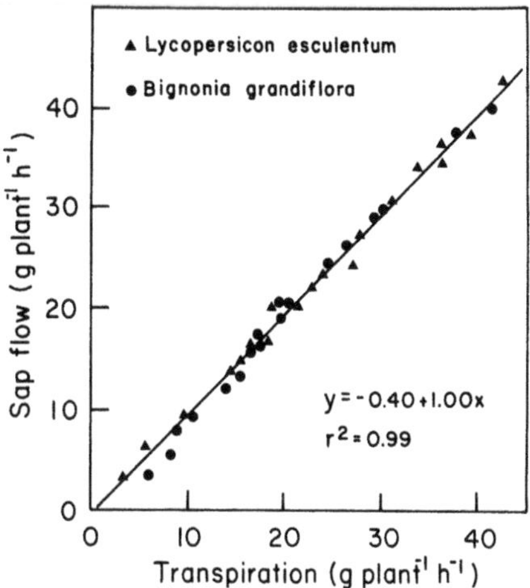

Fig. 8. Calibration of the sap flow rate measured by a heat balance method vs. the transpiration rate determined by weighing for plants of *Lycopersicum esculentum* and *Bignonia grandiflora* with a stem diameter of 10 and 12 mm, respectively. (Fichtner and Schulze 1990)

level by varying the power input into heating, and the temperature measurement sensors are inserted into the stem. This approach was applied recently by Fichtner and Schulze (1990) in a 10-mm stem diameter tropical herbaceous species (*Lycopersicum esculentum* Mill.) and in a 12-mm stem diameter woody species (*Bignonia grandiflora* jacp.). The regression line between sap flow and transpiration measured by weight loss (Fig. 8) indicates a very high accuracy of the measurement.

The accuracy for higher flow rates was tested by letting water flow gravimetrically at a rate ranging from 0 to 2500 h^{-1} through a 17-mm diameter cut stem segment. The relationship between water flow in the section and measured flow by the applied method was the same as that in Fig. 8. The steady-state heat balance method was also tested by Kitano and Eguchi (1989) for cucumber (*Cucumis sativus* L.) with a resolution of 1 mg water loss and a time constant of 1 min.

This high accuracy of the sap flow measurement using the heat balance method with variable heating (null balance) is attributed to both the maintenance of a constant temperature in the stem section and the insertion of the temperature sensors into the stem. Controlling the stem temperature by varying the heat power eliminates potential injury to the stem by overheating, which may occur when the sap flow rate is low, in addition, it eliminates a potential error in sap flow computation when the temperature difference is small at a high sap flow rate (Baker and Nieber 1989; Ishida et al. 1991). Inserting the temperature sensors into the stem can appreciably improve the temperature detection of the moving sap and the xylem tissue. In trees, it has long been recognized that the sap flow rate along the stem cross section is not uniform. Cermak et al. (1973, 1984), using the heat balance method with variable heating power, demonstrated the gradient in sap flow and emphasized the necessity of considering the nonuniformity of flow in different annual rings. Their system device was built to account for this problem by inserting the temperature sensors into the tree trunk. Sakuratani (1981) was also aware of the importance of accurate detection of stem temperature and, in sunflower, he used the surface temperature measurement for stems with a diameters of less than 10 mm, but for larger stems the thermocouples were inserted in the stem. Steinberg et al. (1990b) inserted the thermocouple 2 mm into the bark of a 5-year-old pecan tree in order to

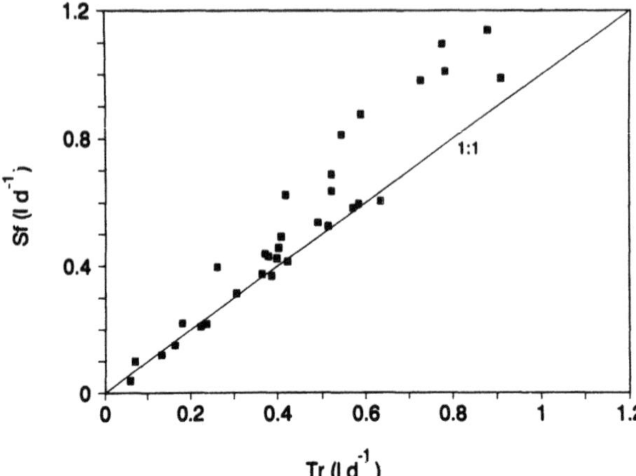

Fig. 9. Daily rate of sap flow in corn as measured by a heat balance method (constant heating) vs. transpiration rate determined by weight loss. (Cohen et al. 1992)

minimize the effect of environmental variations on temperature measurement; however, variation of temperature in the bark may be rather high. Valancogne and Nasr (1989) inserted the thermocouple into the stem and obtained good agreement between calculated sap flow and transpiration in potted trees from a greenhouse having a 10–80-mm stem diameter.

A specific problem regarding the uniformity of the temperature in the cross-sectional area arises when using the method to determine sap flow in monocotyledons. The vascular structure in these species is distributed unevenly (Foster and Gifford 1974), and a simulation of the temperature profile revealed a substantial variation in temperature in the radial direction (Ishida et al. 1991). Due to this variation, the surface temperature does not provide a good representation of the temperature in the cross-sectional area, and calculation of the flow based on surface temperature measurement could lead to a significant error in the estimation of sap flow. Figure 9 presents a set of calibrations of the heat balance method, using the constant power and surface temperature measurement, with potted maize plants in a plastic chamber (32 × 5 m) located in Tottori, Japan (Cohen et al. 1992).

At low sap flow rates, agreement between daily transpiration rate and sap flow was very good, but at high rates, a serious deviation between the two measurements was obtained. This deviation was explained by the nonuniform structure of the conducting elements in the stem which causes temperature differences on the stem surface above and below the heater to be lower than the actual temperature difference in the entire stem area.

The tests described above, using the steady-state heat balance method with variable heating power and constant temperature difference, show that the method is adaptable to both tree and herbaceous species, and that the transpiration rate can be determined at a very high accuracy; however, the measurement requires accurate monitoring of heat supply and temperature detection at high resolution. This should not be a problem in view of newly developed measurement devices.

The verification tests of the heat balance method, the constant or the variable heating power demonstrate the necessity of verifying the method for each species. The assumption that the technique provides an accurate estimation of the mass flow of water and does not need calibration may be incorrect for some species or under certain environmental conditions.

4 Application of the Methods in Research Studies

The tests and verifications of the different methods presented in the previous section show that sap flow measurement can provide a good estimation of transpiration (Tr) from a whole plant. The sap flow measurement may be used as a link between leaf studies and micrometeorological techniques above canopies. In field studies, when investigating the plant–water relationship or when attempting to improve the irrigation management of crops, Tr from the canopy, rather than Tr from a single plant, is required; therefore, it is important that samples of several plants be used to extrapolate a value for the entire canopy. Under field conditions, variability among plants may be high due to environment and soil variations and Tr must be determined for a sufficient number of plants to satisfy statistically the requirement for an accurate representation of the entire canopy. The size of the sample is a function of variability among plants, but under field conditions and particularly in studies of trees, it is difficult to determine a criterion for variability among individuals.

The most convenient criterion considered in different studies has been diameter of the stems, although for some species, this parameter may not be representative of leaf area or Tr. In a pine stand, no relationship was found between sap flow and circumference of the trunk. It was suggested that the arithmetic mean of a representative sample can be regarded as the best estimation of mean sap flow (Diawara et al. 1991). In this study, a sample of ten trees was required to provide an estimate of the mean sap flow with a confidence interval $< 10\%$.

In a 300 m^2 stand of spruce trees, the sap flow rate was determined for 14 trees whose stem area represented all tree sizes in the plot (Cermak and Kucera 1987). Selection of sampled trees in a citrus orchard was also based on stem size. In this study, the selected diameter range reflects the distribution of the stem diameters in the entire experimental plot (Cohen 1991). The standard deviation of Tr, determined after the second year of the experiment on 22 sampled trees (Fig. 10), indicates that the number of sampled trees must be high in order to obtain the orchard Tr with reasonable accuracy. For example, a sample of 15 trees was required in order to estimate the orchard Tr with an error of 10% of the mean.

Stem size was also used as a criterion for plant selection in herbaceous species. Petersen et al. (1992) selected ten cotton plants representative of the full range of stem diameters found in the field. A similar selection of eight to ten plants was made by Ham et al. (1990, 1991). Sap flow measurements from individual plants were extrapolated for canopy Tr by normalizing the data on a leaf area basis. Both studies presented very good agreement between canopy Tr determined by the sap flow and other independent measurements of Tr. But Dugas (1990a) showed that using the stem diameter as a criterion for selecting a representative sample of the canopy may introduce a significant error in the estimation of canopy Tr. Sap flow for the sampled plants in a lysimeter was consistently greater (by nearly 25%) than the lysimeter Tr. The relationship between daily Tr from lysimeter and sap flow is shown in Fig. 11.

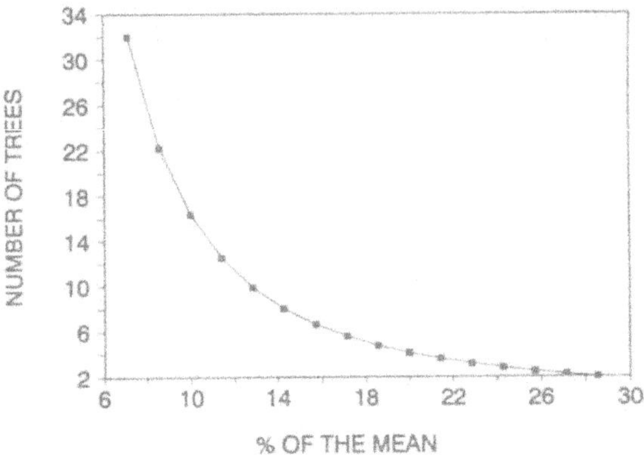

Fig. 10. Number of citrus trees required (at a 95% confidence limit) to estimate orchard transpiration to within a given deviation from the mean sample. (Cohen 1991)

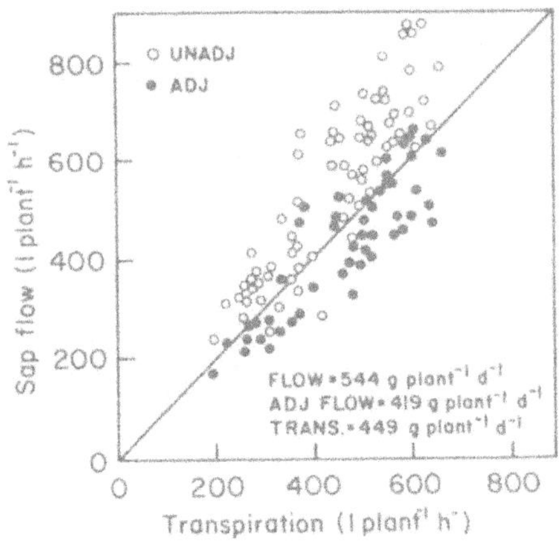

Fig. 11. Daylight totals of cotton sap flow rate measured by the constant heating power, adjusted (*ADJ*) and unadjusted (*UNADJ*) for stem area differences vs. lysimeter transpiration. (Dugas 1990a)

Adjustment was made to correct for stem area, as the stem area of plants with gauges was larger than the average stem area of lysimeter plants. Although plants close in size had been selected, the midday hourly values of sap flow ranged from 30 to 100 $g\,h^{-1}\,plant^{-1}$.

Sap flow measurement has been applied in several field studies to determine canopy Tr. Tr was related generally to potential evapotranspiration (PTr), calculated from meteorological variables by the Penman-Monteith equation. In a fully grown spruce (*Picea abie* L. karst) stand located in southern Moravia, Tr determined by sap flow and PTr were linearly related. A simple empirical model for estimating Tr from the stand under nonlimiting soil moisture conditions showed the ratio Tr/PTr to be 0.57 for the spruce stand (Cermak and Kucera 1987). In a stand of maritime pine (*Pinus pinaster* Ait.) located in southwest France, the relationship between daily Tr from the stand measured by the sap flow and PTr for 36 days was 0.55 (Granier et al. 1990). A very similar ratio (0.52) was also reported by Cohen (1992) from Tr data determined by the sap flow rate in eight pine trees (*Pinus halepensis* cv. Mill.) located in the northern part of Israel. Using the same technique for sap flow measurement as in pine trees, Granier (1987) reported a similar Tr/PTr ratio (0.6) for a Douglas-fir [(*Pseudotsuga menziesii* (Mirb.) Franco] stand from measurement of stand sap flow and PTr using the Penman model. The relationship between Tr and PTr was linear (Tr = − 0.11 + 0.6 PTr). When available soil water decreased below 30% of its maximum value, transpiration fell below its maximum. Diawara et al. (1991) used sap flow measurements in pine trees to estimate the proportion of the Tr from the trees in relation to the evapotranspiration from the stand including the understorey vegetation. Sap flow measurement was used for corn to estimate the effect of watering various proportions of the root system on a plant–water status (Gavloski et al. 1992).

Dugas et al. (1992) used the sap flow measurement to assess the biological and hydrological implications of the increasing density of honey mesquite (*Prosopis glandulosa* Torr. var. glandulosa) on watersheds of the southwestern rangelands in the USA. In these species, as in the tree species mentioned above, the diurnal pattern of Tr was linear to that of PTr. A linear relationship between Tr and PTr was also reported by Edwards and Warwick (1984) for a kiwifruit vine. Sap flow measurement in the trunk and PTr were used during two irrigation seasons to determine the water requirement of a citrus orchard. Tr was proportional to PTr, and the ratio between the two, averaged for 22 trees, was approximately 0.3 when soil water availability was not limited (Cohen 1991).

The ability to accurately measure the Tr from crops by sap flow measurement was used by several workers to verify models that computed the partitioning of energy balance above crops. Sakuratani (1987) compared the sap flow data from a soybean canopy using the Bowen ratio heat balance method. This comparison allowed an indirect estimation of the evaporation from a soil surface, which was related to the leaf area index. A similar application of the method in combination with the Bowen ratio energy balance was made in cotton (cv. Paymaster 404) to study the partitioning of evapotranspiration during a period of partial cover (Ham et al. 1990, 1991). Their study indicated that the use of directly measured latent heat flux from the canopy can eliminate the need for simplifying assumptions about energy transfer in the estimation of energy exchange within the soil-canopy-atmosphere system. The sap flow measurement was applied also in cotton (cf. Acala SJ-2) to validate a model computing Tr from sunlit and shaded fractions of the canopy using the Penman-Monteith energy balance equation (Petersen et al. 1992). Figure 12 shows the linear regression between hourly computed Tr by means of the energy balance equation and sap flow for 23 days. The R^2 value of computed Tr for sap flow is 0.96 and the standard error of the prediction is 0.06 $mm\,h^{-1}$.

Sap flow measurements have been used to study the contribution of water storage in trees to Tr. The measurement is generally made near the stem base, and in herbaceous species sap flow values are generally identical or very close to Tr from the canopy. But for many tree species, the diurnal curve of the sap flow generally lags behind that of Tr. The difference between the pattern of the two curves in trees is used to estimate water storage in the tree. Using the sap flow measurement, Cermak et al. (1984) calculated the amount of water storage in crack willow tree (*Salix fragilis* L.) as 3% of Tr during a summer day or nearly 4 $kg\,day^{-1}$. Estimation of water storage was also made for intact, naturally growing adult *Larix decidua* L. and *Picea abies* trees (Schulze et al. 1985). They found the total amount of available water storage to be 24 and 14% of the total daily Tr in *Larix* and *Picea*, respectively. Using

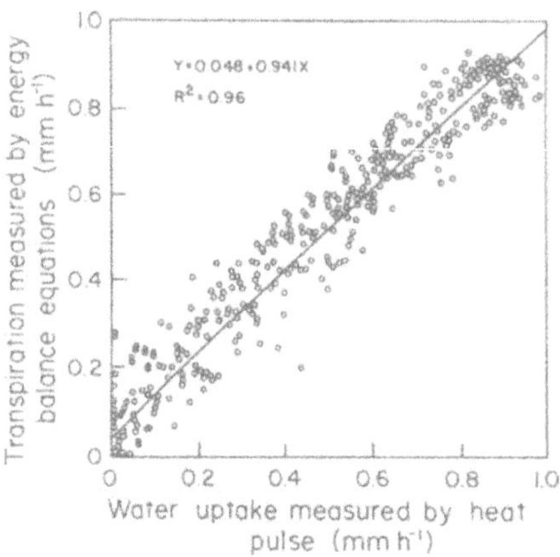

Fig. 12. Hourly computed transpiration measured by energy balance equations vs. hourly water uptake of cotton measured by the heat pulse method for 23 days spanning nonstress to severe stress conditions. (Petersen et al. 1992)

a very sensitive sap flow measurement, Kitano and Eguchi (1989) showed that in cucumber, water absorption by the root lagged about 10 min behind leaf Tr. Temporary water loss caused by a rapid increase in canopy Tr was buffered by about 5% of the water content in the stem.

Sap flow measurement has proven to be a very useful tool in studies of water transport in trees. Measurements may be done in the trunk to provide the water flow rate for the entire tree, but at the same time they may be performed in the branches to determine the distribution of water flow in different parts of the tree. This measurement was generally combined with measurements of soil and representative leaf water potential in order to study the driving force of water transport in the system. This approach was applied to a 22-year-old citrus tree to study the effect of soil water stress on the hydraulic conductance of the tree. Hydraulic conductance was computed from the relationship between the diurnal curves of hourly values of sap flow and leaf water potential (Cohen et al. 1983). A similar approach was applied to a 5-year-old pecan tree (Steinberg et al. 1990b). Study of the partitioning of hydraulic conductance in mature citrus trees (cv. Valencia and Shamouti), using the sap flow and leaf water potential measurements, was made by Moreshet et al. (1990). They computed the hydraulic conductance of the transpiring crown from sap flow and the weighted average of sunlit and shaded leaf water potential. Root system conductance was estimated from sap flow and covered leaf water potential.

Application of the sap flow measurement to irrigation practices has been attempted in citrus orchards. The effect of partial wetting of the soil, using minisprayers, compared with the wetting of the entire soil volume, on water use by the trees, was examined by Moreshet et al. (1983). An attempt to use the method as a criterion for irrigation scheduling in citrus was made by Cohen (1991) during two irrigation seasons. The change in Tr/PTr during the irrigation intervals was used to determine orchard water use.

Tr/PTr on the first day after irrigation (Tri/PTri) and on the last day of the irrigation interval (Tre/PTre) are given in Fig. 13. The difference between the two lines indicates the extent to which Tr dropped due to the reduction in soil water availability. Irrigation was resumed when Tre/PTre was about 80% of Tri/PTri.

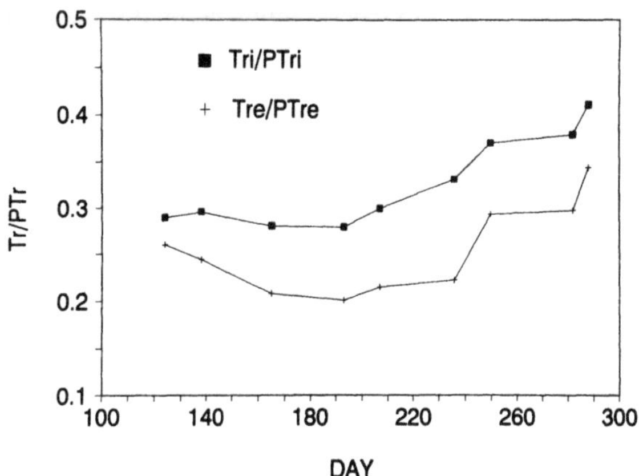

Fig. 13. Ratio of measured to potential transpiration at the beginning of the irrigation interval (Tri/PTri) and at the end of the interval (Tre/PTre). Nine intervals are shown throughout the irrigation season in 1988. (Cohen 1991)

5 Conclusions

The growing interest in sap flow measurement throughout the world indicates that this method has been widely recognized as a reliable measure of transpiration in plants. Enhanced development of system devices for sap flow measurement over the past two decades has been the main reason for the substantial progress made in applying the method to different species. The method has already been proved by a large number of studies to be a valuable tool, and perhaps the most reliable one, for investigating different bioclimatological aspects of water loss in the field. However, the theory of heat dissipation in the stem is still based on simplifying assumptions and provides an unsatisfactory solution to the different approaches. The assumptions used in the models are related to the structure and function of the water-conducting system in the stem, which varies considerably between species. For this reason, a test or verification of the models is required for each species when either the heat pulse or the steady-state heat balance methods is applied.

For a single plant, the transpiration rate may be estimated by the sap flow measurement with an accuracy of 10% or better. Selection of the method should be based on plant characteristics and on the objectives of the study. Stem diameter, general knowledge of the structure of the water-conducting system in the stem, and the expected maximum transpiration rate are the most important plant criteria for selecting the measurement method. Other criteria, not related to the plant, are the sensitivity of the method to environmental variations, convenience in installing the sensors on the plant, and simplicity of the measurement device. These latter criteria are particularly relevant when the study is conducted under field conditions.

Application of the method in several fields studies has shown that sap flow measurement is capable of providing a good estimation of the canopy transpiration rate. For this purpose the integrated sap flow values of several selected plants are required. The expected accuracy of the measurement is obviously less than for a single plant, as the variability among plants contributes an additional source of error. The major problem in this integration technique is a reliable criterion of the variability among plants for accurate determination of sample size.

The ability to successfully estimate the canopy transpiration in trees and herbaceous species suggests that the sap flow method can be used potentially to improve irrigation management and plant water use. The main advantage of this approach is that application of water in the field will be based principally on functions describing the crop's response to actual water uptake; however, the present system devices, both those built in individual laboratories or those available commercially, are unsuitable for farm use. Further development of measurement systems to substantially simplify the operational procedure and data analysis may enhance the use of this method for irrigation practice.

The sap flow method can be used in research on hydrology of irrigated areas and ground-water pollution by fertilizers and pesticides. The amount of drained irrigation water to the ground below the root zone can be controlled due to the ability to accurately estimate the crop water requirement. The importance of this aspect is gradually increasing in many regions in the world where good water quality sources are declining and irrigation with low water quality is inevitable. Accurate knowledge of the water flow rate through the plant would allow a better understanding of the soil-plant-atmosphere relationship, consequently, irrigation and fertilization management as well as knowledge of plant response to climatic conditions would be considerably improved.

References

Baker JM, Nieber JL (1989) An analysis of the steady-state heat balance method for measuring sap flow in plants. Agric For Meteorol 48: 93–110

Baker JM, van Bavel CHM (1987) Measurement of mass flow of water in the stems of herbaceous plants. Plant Cell Environ 10: 777–782

Bloodworth EM, Page JB, Cowley WR (1955) A thermoelectric method for determining the rate of water movement in plants. Soil Sci Soc Am Proc 19: 411–414

Cermak J, Kucera J (1987) Transpiration of mature stands of spruce (*Picea abies* L. Karst.) as estimated by the tree-trunk heat balance method. In: Forest hydrology and watershed management. Proc Vancouver Symp, Aug 1987. IAHS Publ 167, pp 311–317

Cermak J, Deml M, Penka M (1973) A new method of sap flow rate determination in trees. Biol Plant 15: 171–178

Cermak J, Ulehla J, Kucera J, Penka M (1982) Sap flow rate and transpiration in the full-grown oak (*Quercus robus* L.) in floodplain forest exposed to seasonal floods as related to potential evapotranspiration and tree dimensions. Biol Plant (Praha) 24: 446–460

Cermak J, Jenik J, Kucera J, Zidek V (1984) Xylem water flow in a crack willow tree (*Salix fragilis* L.) in relation to diurnal changes of environment. Oecologia 64: 145–151

Closs RL (1958) Heat pulse method for measuring rate of sap flow in a plant stem. N Z J Sci 1: 281–288

Cohen Y (1991) Determination of orchard water requirement by a combined trunk sap flow and meteorological approach. Irrig Sci 12: 93–98

Cohen Y (1992) A combined transpiration measurement and meteorological approach for determining water requirement in the field and orchard. In: Shalhevet J, Changming L, Yuexian X (eds) Water use efficiency in agriculture, Proc of the Binational China-Israel Worksh, April 22–26, 1991 Beijing, China, Priel Publishers, Rehovot, pp 170–180

Cohen Y, Fuchs M (1989) Problems in calibrating the heat pulse method for measuring sap flow in the stem of trees and herbaceous plants. Agronomie 9: 321–325

Cohen Y, Fuchs M, Green GC (1981) Improvement of the heat pulse method for determining sap flow in trees. Plant Cell Environ 4: 391–397

Cohen Y, Fuchs M, Cohen S (1983) Resistance to water uptake in mature citrus tree. J Exp Bot 34: 451–460

Cohen Y, Kelliher FM, Black TA (1985) Determination of sap flow in Douglas-fir trees using the heat pulse technique. Can J For Res 15: 422–428

Cohen Y, Fuchs M, Falkenflug V, Moreshet S (1988) Calibrated heat pulse method for determining water uptake in cotton. Agron J 80: 398–402

Cohen Y, Huck MG, Hesketh JD, Frederick JU (1990) Sap flow in the stem of water stressed soyabeans and maize plants. Irrig Sci 11: 45–50

Cohen Y, Takeuchi S, Nozaka J, Yano T (1992) Accuracy of sap flow measurements in herbaceous plants using heat balance or heat pulse methods. Agron J (in press)

Daum CR (1967) A method for determining water transport in trees. Ecology 48: 425–431

Diawara A, Loustau D, Berbigier P (1991) Comparison of two methods for estimating the evaporation of a *Pinus pinaster* (Ait.) stand: sap flow and energy balance with sensible heat flux measurements by an eddy covariance method. Agric For Meteorol 54: 49–66

Doley D, Grieve BJ (1966) Measurement of sap flow in a eucalyptus by thermoelectric method. Aust For Res 2: 3–27

Dugas WA (1990a) Comparative measurement of stem flow and transpiration in cotton. Theor Appl Climatol 42: 215–221

Dugas WA (1990b) Sap flow in stems. In: Francois B (ed) Remote sensing reviews. Harwood Academic Publishers, New York, pp 225–235

Dugas WA, Marcus LH, Herman SM (1992) Diurnal measurements of honey mesquite transpiration using stem flow gauges. J Range Manage 45: 99–102

Edwards WRN, Warwick NWM (1984) Transpiration from a kiwifruit vine as estimated by the heat pulse technique and the Penman-Monteith equation. N Z J Agric Res 27: 537–543

Fichtner K, Schulze ED (1990) Xylem water flow in tropical vines as measured by a steady state heating method. Oecologia 82: 355–361

Foster RD, Gifford EM (1974) Comparative morphology of vascular plants. Freeman, San Francisco, 751 pp

Gavloski JE, Whitfield GH, Ellis CR (1992) Effect of restricted watering on sap flow and growth in corn (*Zea mays* L.). Can J Plant Sci 72: 361–368

Granier A (1985) Une nouvelle methode pour la mesure du flux de seve brute dans le tronc des arbres. Ann Sci For 42: 81–88

Granier A (1987) Evaluation of transpiration in a Douglas-fir stand by means of sap flow measurements. Tree Physiol 3: 309–320

Granier A, Bobay V, Gash JHC, Gelpe J, Saugier B, Shuttleworth WJ (1990) Vapor flux density and transpiration rate comparisons in a stand of maritime pine (*Pinus pinaster* Ait.) in les landes forest. Agric For Meteorol 51: 309–319

Green SR, Clothier BE (1988) Water use of kiwifruit vines and apple trees by the heat-pulse technique. J Exp Bot 39: 115–123

Groot A, King KM (1992) Measurement of sap flow by the heat balance method: numerical analysis and application to coniferous seedlings. Agric For Meteorol 59: 289–308

Ham JM, Heilman JL (1990) Dynamics of a heat balance stem flow gauge during high flow. Agron J 82: 147–152

Ham JM, Heilman JL, Lascano RJ (1990) Determination of soil water evaporation and transpiration from energy balance and stem flow measurements. Agric For Meteorol 52: 287–301

Ham JM, Heilman JL, Lascano RJ (1991) Soil and canopy energy balance of a row crop at partial cover. Agron J 83: 744–753

Hatton TJ, Catchpole EA, Vertessy RA (1990) Integration of sap flow velocity to estimate plant water use. Tree Physiol 6: 201–209

Huber B (1932) Beobachtung und Messung Pflänzlicher Saftstrome. Ber Dtsch Bot Ges 50: 89–109

Huber B, Schmidt E (1937) Eine Kompensationsmethode zur thermoelektrischen Messung langsamer Saftstrome. Ber Dtsch Bot Ges 55: 514–529

Ishida T, Campbell GS, Calissendorff C (1991) Improved heat balance method for determining sap flow rate. Agric For Meteorol 56: 35–48

Kitano M, Eguchi H (1989) Quantitative analysis of transpiration stream dynamics in an intact cucumber stem by a heat flux control method. Plant Physiol 89: 643–647

Kozlowski TT, Leyton L, Hughes JF (1965) Pathways of water movement in young conifers. Nature 205: 830–832

Kucera J, Cermak J, Penka M (1977) Improved thermal method of continually recording the transpiration flow rate dynamics. Biol Plant 19: 413–420

Ladefoged K (1960) A method for measuring the water consumption of larger intact trees. Physiol Plant 13: 648–658

Lassoie JP, Scott DRM, Fritschen LJ (1977) Transpiration studies in Douglas-fir using the heat pulse technique. Forest Sci 23: 377–390

Leyton L (1970) Problems and techniques in measuring transpiration from trees. In: Luckwill LC, Cuting CV (eds) Physiology of tree crops. Proc of 2nd Long Ashton Symp, 25–28 March 1969. Academic Press, London, pp 101–111

Mark WR, Crews DI (1973) Heat-pulse velocity and bordered pit condition in living Engelmann Spruce. Forest Sci 19: 291–296

Marshall DC (1958) Measurement of sap flow in conifers by heat transport. Plant Physiol 33: 385–396

Miller DR, Vavrina CA, Christensen TW (1980) Measurement of sap flow and transpiration in ring-porous oaks using a heat pulse velocity technique. For Sci 26: 485–494

Moreshet S, Cohen Y, Fuchs M (1983) Response of mature Shamouti orange trees to irrigation of different soil volumes at similar levels of available water. Irrig Sci 3: 223–236

Moreshet S, Cohen Y, Green GC, Fuchs M (1990) The partitioning of hydraulic conductance within mature orange trees. J Exp Bot 41: 833–839

Morikawa Y (1972) The heat pulse method and an apparatus for measuring sap flow in woody plants. J Jpn For Sci 54: 166–171

Petersen KL, Fuchs M, Moreshet S, Cohen Y, Sinoquet H (1992) Computing transpiration of sunlit and shaded cotton foliage under variable water stress. Agron J 84: 91–97

Pickard WF (1973) A heat pulse method of measuring water flux in woody plant stems. Math Biosci 16: 247–262

Sakuratani T (1981) A heat balance method for measuring water flux in the stem of intact plants. J Agric Meteorol 37: 9–17

Sakuratani T (1984) Improvement of the probe for measuring water flow rate in intact plants with the stem heat balance method. J Agric Meteorol 40: 273–277

Sakuratani T (1987) Studies on evapotranspiration from crops (2). Separate estimation of transpiration and evaporation from a soyabean field without water shortage. J Agric Meteorol 42: 309–317

Sakuratani T (1990) Measurement of the sap flow rate in stem of rice plant. J Agric Meteorol 45: 277–280

Schulze ED, Cermak J, Matyssek R, Penka M, Zimmerman R, Vasicek F, Gries W, Kucera J (1985) Canopy transpiration and water fluxes in the xylem of the trunk of *Lariz* and *Picea* trees – a comparison of xylem flow, porometer and cuvette measurements. Oecologia 66: 475–483

Shakel RA, Johnson RS, Medawar CK, Phene CJ (1992) Substantial errors in estimates of sap flow using the heat balance technique on woody stems under field conditions. J Am Soc Hort Sci 117: 351–356

Steinberg SL, van Bavel CHM, McFarland MJ (1989) A gauge to measure the mass flow of sap in stems and trunks of woody plants. J Am Soc Hort Sci 114: 466–472

Steinberg SL, van Bavel CHM, McFarland MJ (1990a) Improved sap flow gauge for woody and herbaceous plants. Agron J 82: 851–854

Steinberg SL, McFarland MJ, Worthington JW (1990b) Comparison of trunk and branch sap flow with canopy transpiration in pecan. J Exp Bot 41: 653–659

Stone JF, Shirazi GA (1975) On the heat-pulse method for the measurement of apparent sap velocity in stems. Planta 122: 169–177

Swanson RH (1972) Water transpired by trees is indicated by heat pulse velocity. Agric Meteorol 10: 277–281

Swanson RH (1983) Numerical and experimental analysis of implanted-probe heat pulse velocity theory. PhD Thesis, University of Alberta, Edmonton, Alberta, Canada

Swanson RH, Whitfield WA (1981) A numerical analysis of heat pulse velocity theory and practice. J Exp Bot 32: 221–239

Valancogne C, Nasr Z (1989) Une methode de measure du debit de seve brute dans petits arbres par bilan de chaleur. Agronomie 9: 609–617

Vieweg GH, Ziegler H (1960) Thermoelektrische Registrierung der Geschwindigkeit des Transpirationsstromes. Ber Dtsch Bot Ges 73: 221–226

Wendt CW, Brooks CR, Runkles JR (1965) Use of the thermoelectric method to measure relative sap flow in monocotyledons. Agron J 57: 637–638

4 Laser Remote Sensing of Vegetation

V.A. Kanevski, J. Ross, S.M. Kochubey, and T. Shadchina

1 Introduction

Most of the presently available methods for remote sensing of plant growth and development are based on measurements of spectral reflectance factors (Ross 1981; Kondratyev and Fedchenko 1982; Rachkulik and Sitnikova 1981; Bauer 1985; Shibayama and Akiyama 1986). These characteristics, however, contain information about the plants themselves, as well as the singularities of the solar radiation regime and other physical parameters of the environment (Kondratyev and Fedchenko 1982). It is, therefore, very difficult to find an unambiguous relationship between the signals measured and parameters that characterize the vegetation condition, such as biomass, chlorophyll and moisture content, occurrence of mineral deficiency, and the presence of other stresses. The procedure for measuring reflectance factors is concerned with the need to measure the standard reflecting surfaces which receive the same illumination as the object. This complicates the measurements considerably and appreciably decreases the accuracy of the reflectance factors measured.

To overcome these difficulties, new methods of remote sensing have recently been advanced and are now being intensively developed (Kondratyev et al. 1987; Kochubey et al. 1990). One of them is a method of passive probing using airborne spectrometers of high spectral resolution. These instruments make it possible to measure the spectral distribution of intensity in a reflected light flux with a resolution of 1 to 10 nm (Belyaev et al. 1978; Collins 1978). Remote and laboratory measurements have shown that the "red edge" of the reflection spectrum depends on the chlorophyll concentration in plant leaves (Horler et al. 1983). Our studies have developed methods to determine the chlorophyll and total nitrogen content of plant leaves that use quantitative parameters similar to the reflection spectrum in the "red edge" region (Kochubey et al. 1987, 1990). Since the form of the spectrum depends only on the relative distribution of intensity in a narrow spectral interval of 680 to 750 nm, our method does not require measurements of reflectance factors, that is, it is not necessary to normalize the reflection spectrum to a standard. This increases the accuracy of measurements and makes the method more effective than those in wide use now. Successful application of the method confirms the prediction that high-resolution spectroscopy could be useful in remote sensing in the near future (Ferns et al. 1984). Extensive development of instruments such as videospectrometers will accommodate measurements of this type to space research.

Another new direction in remote sensing research is active probing in the optical range with the use of lasers. The presently available experimental data suggest that studies of laser-induced, plant luminescence emission, as well as of the spatial structure and time parameters of scattered secondary light flux can be most promising. The possibility of concentrating laser light energy in a narrow beam allows one to identify, through remote sensing, plant canopies with a high spatial resolution, making it possible to distinguish individual plants and even individual leaves. Thus, probing using the pulse mode of laser operation with a 1–10 picosecond light pulse enables one to measure the vertical structure of the plant canopy with high resolution. The plant luminescence parameters provide important information about the condition of a plant. This information can be derived by applying some methods developed in laboratory studies (Karnaukhov 1978; Kochubey et al. 1986, 1990; Lichtenthaler 1987, 1988a; Lichtenthaler and Rinderle 1988; Rinderle and Lichtenthaler, 1988). Remote measurements are specific in that they rely on a limited choice of parameters which must be measured accurately. It is, therefore, necessary to understand clearly what information can be derived from such measurements.

The aim of the present review is to generalize the available data concerning the use of lasers in remote sensing research. An attempt is made to analyze the specific features, as well as the informative potential and prospects of such research, showing how laser technique fits into the general framework of remote sensing.

2 Remote Laser Spectrofluorometry of Vegetation

Classical optical methods of studying the state of an object involve the analysis of different types of secondary radiation. These, in addition to reflected and scattered light flows, include luminescence. Analysis of the parameters of luminescence makes it possible to obtain information concerning the object that produces luminescence measurable at a distance. This method is effective in studies of plants, because it is based on the inertialess nature of fluorescent emission, is highly sensitive, and enables experimentation with undisturbed objects. The advantages of fluorescence studies of plant leaves, subcellular organelles, and suborganelle particles were shown by Kochubey (1986) and Lichtenthaler and Rinderle (1988). The development of the laser technique and the advances in designing opto-electronic systems have created the necessary technical prerequisites for luminescence methods to be used in remote investigations of plants. Remote measurements impose certain restrictions on the choice of parameters of luminescence emission that can be measured reliably using the instruments available now. It is, therefore, necessary to discuss what information can be obtained and what information is beyond the present capabilities of measuring instruments.

2.1 Theoretical Foundations of Fluorescence Sensing Methods

The basic parameters that characterize the luminescence of substances are the luminescence intensity, the intensity distribution over wavelengths or the luminescence spectrum, the decay time, and the polarization of this emission. The measurements of chlorophyll fluorescence induction kinetics are quite informative and widely used for plant leaves. The theory and the potential of the method were discussed by many authors (Karapetyan and Bukhov 1986; Lichtenthaler 1988a). Portable fluorimeters make it possible to use the induction method to examine the state of plants in field crops, but it is very difficult to employ this method for remote measurements.

The intensity and the spectrum of luminescence are practically the only parameters of luminescence emission used presently in remote measurements. Measurements of the distribution of luminescence decay over wavelengths have been reported (Measures et al. 1974). Since it is very difficult to calibrate radiation detectors in absolute units of energy, brightness, etc., the intensity of luminescence is usually measured with reference to some standard source of radiation. In aerial remote measurements the combination scattering lines of atmospheric nitrogen may serve as a standard (Bunkin et al. 1984). Measurements of the spectral forms allow relative parameters such as the ratios of intensities at different wavelengths to be obtained. It will be shown below that these parameters provide information about the physiological state of plants. Moreover, they are stable for variations under experimental conditions. Modern instruments make it possible to measure these parameters simply and reliably. We think that these parameters are very promising for remote measurements.

The luminescence spectrum of green plant leaves in the visible range involves two bands (685–690 and 735–740 nm) in the red region, and two wider, weakly structured bands in the blue-green region, with the maxima at about 440–480 and 510–540 nm. The "red" luminescence is emitted by chlorophyll *a* and is fluorescence physically. The contribution of fluorescent emission by chlorophyll *b* to the "red" luminescence of plant leaves is vanishingly small because the time necessary for the energy transfer from chl. *b* to chl. *a* is much less than the lifetime of the excited state of the chl. *a* molecule. The effective energy exchange is also observed between chl. *a* molecules, resulting in almost a tenfold decreased quantum yield of fluorescence, defined as the ratio of emitted to absorbed light quanta. The quantum yield of the "red" fluorescence of plant leaves is generally less than 5%. For chl. *a* in photosynthetic membranes, the position and the form of the two main fluorescence bands, their relative intensity, and origin are different from the same characteristics of the spectrum of its solutions. The long-wavelength shift of the first band maxima and its broadening are primarily related to the specific organization of a pigment system. It is known to be represented by a set of chl. *a* molecule aggregates, each of which has its own absorption band. The maxima of the appropriate fluorescence bands are in the range 670–700 nm (Kochubey 1986). The 685-nm fluorescence band of the leaves is formed by the superposition of the short-wavelength fluorescence bands of various chlorophyll aggregates. Their vibrational satellites, as well as the emission by minor antenna chlorphyll forms, are in the region 710–750 nm. They generate the second band in the fluorescence spectrum of leaves. The contribution from the antenna forms to long-wavelength emission at room temperature is very small compared to the luminescence of vibrational satellites (Tusov et al. 1980). When the photosynthetic apparatus becomes damaged, the luminescence yield of antenna forms can increase (Lichtenthaler and Rinderle 1988); thus, it rises very markedly when plant leaves are cooled to low temperatures. At 77 K the 735–740-nm fluorescence is due entirely to the emission of antenna forms (Kochubey 1986).

In the fluorescence spectrum of dark-green leaves the 735–740 nm band is more intensive than the 685–690 nm band. This is not due to the increased fluorescence yield of long-wavelength forms, as suggested by some researchers, but due to the specific optical properties of a leaf tissue which has a heterogeneous optical system with different degrees of ordering of its elements. The chloroplasts containing chlorophyll have a volume much below that of the cells in which they are irregularly distributed. The chlorophyll is connected with the chloroplast membranes which are distributed in a nonuniform manner inside the chloroplast and are embedded in a colorless stroma. The increased concentration of a pigment in some places and its absence in others, as well as the insertion of chlorophyll carriers into a strongly dispersive medium, are the reasons for the highly distorted absorption and fluorescence spectra of chloroplasts and leaves compared to those of solutions. The additional distortion of leaf spectra is induced by the nonuniform distribution of chloroplasts in the plant cells and tissues. Thus, the reabsorption effect is much enhanced, and it is possible to show that this effect influences the intensity ratio of the 685- and 735-nm bands. The reabsorption is due to the fact that the 685-nm fluorescence band overlaps the absorption band considerably. An increase in the optical path length in a dispersive medium results in enhanced absorption, so that fluorescence emission is absorbed at the short-wavelength edge of the 685-nm band. This results not only in a decrease in its intensity, but also in an apparent long-wavelength shift of its maximum. Our measurements of the chloroplast fluorescence spectra with a gradually decreasing chlorophyll concentration in a suspension show that beginning with 10% absorption in the region of the red maximum, further dilution leads to the absence of a short-wavelength shift of the 685-nm band maximum and produces no change in the I_{685}/I_{735} ratio. Consequently, the reabsorption effect under the absorption considered is minimal. Similar data are reported by Chappelle et al. (1984) and Lichtenthaler and Rinderle (1988). This specific feature of the fluorescence spectra of chl. *a* in plant leaves can be used as a basis to develop a method of determining its concentration in leaves (Kochubey et al. 1986; Lichtenthaler 1987). Figure 1 shows the I_{685}/I_{735} ratio as a function of pigment concentration in winter wheat leaves. The dependence is described by the following expression (Kochubey et al. 1986):

$$[c]_{chl} = \frac{0.58}{I_{685}/I_{735}} + 0.55.$$

A similar plot is given by Lichtenthaler and Rinderle (1988) for beech leaves. The correlation coefficient determined by us for the above regression equation is 0.84 $\pm$ 0.04. To construct the regression, we used the data from measurements of leaves of varying ages; the leaves were taken from plants grown with different nitrogen supplies. The results showed that the state of the leaf tissue changed, and the amounts of dry matter, the water content of the leaves, and other anatomical parameters differed. The high correlation of the I_{685}/I_{735} ratio with the chlorophyll content confirms that this content is the basic factor that determines the value of the ratio by switching on the reabsorption mechanism. Moreover, the data obtained show that our regression equation enables the chlorophyll content in leaves to be determined irrespective of the period of vegetation and growth environment.

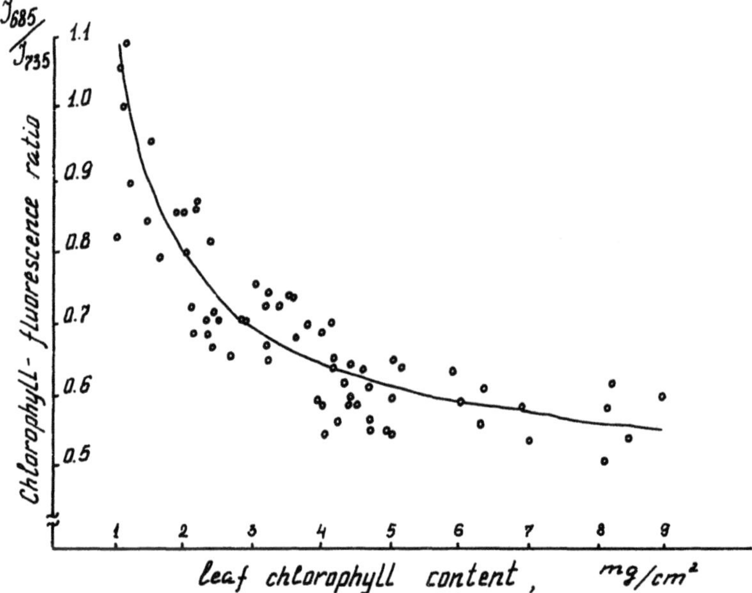

Fig. 1. Dependence of the values of the chlorophyll fluorescence ratio, I_{685}/I_{735}, on the chlorophyll content of wheat leaves. The plants were grown on different levels of nitrogen nutrition. (Kochubey et al. 1986)

In the case of green plant leaves the light falling on the leaf does not penetrate deeply into the leaf tissue as it is almost fully absorbed in surface layers. Ninety-nine percent of the light flux falling on a leaf containing 5 mg chlorophyll/mg moisture content is absorbed by the upper 0.002 nm thickness. It is clear, therfore, that the ratio I_{685}/I_{735} as a function of the chl. *a* content is little affected by the anatomical peculiarities of the leaf tissue structure. It, therefore, appears to be unrealistic to use this relationship to identify plants taxonomically (Chappelle et al. 1985). We thus conclude that by measuring the form of the "red" fluorescence spectrum for plant leaves, that is, the ratio of intensities at the maxima of two main bands, 685 and 735 nm, it is possible to obtain reliable information only on the chlorophyll content in the leaves of any plant. As will be shown below, data concerning the level of plant pigmentation may serve as a basis for methods distinguishing the condition of plant cover.

It is now widely accepted that blue luminescence is emitted by the cellular walls. However, there are some data showing that blue luminescence is also emitted by the cellular components taking part in plant metabolism, e.g., some products of secondary synthesis such as alkaloids, anthraquinones, and flavonoids. The latter are contained only in plants and are polyphenols (Brayon 1966; Blazej and Suty 1973). The phenolcarbonic acids and esters observed in plant cells at a high concentration are also polyphenol compounds. Plant phenol product are metabolically active and are significantly changed by variations in the environment (Barz 1977).

In plant cells important components of the energy exchange systems such as pyridine nucleotides and flavoproteins, can be found. They are associated with various energy-producing processes of the cell such as glycolysis, pentose phosphate and Krebs cycles, fatty acid oxidation, and some paths of terminal oxidation.

A specific feature of pyridine nucleotides and flavoproteins is that their redox transformations are accompanied by appreciable changes in their luminescence yield in the blue region and in the positions of the maxima of relevant bands. Thus, the first compounds fluoresce only in reduced, and the second compounds only in oxidized states. When pyridine nucleotides are bound to dehydrogenases, the maximum of their fluorescence shifts toward shorter wavelengths (about 440 nm) (Lehninger 1972). The derivatives of riboflavine-flavin mononucleotide (FMN) and flavin adenine dinucleotide (FAD) are prosthetic groups of many components of the terminal oxidation systems. The addition of FMN and FAD to a ferment molecule leads to a decrease in the luminescence intensity in a number of cases.

The universal character of the energy processes in plant cells involving pyridine nucleotides and flavoproteins, as well as the specific changes in their fluorescence yield under redox transformations, make it possible to treat these substances as characteristic labels for the activity of the energy apparatus (Duysens and Amez 1957; Estabrook 1962; Kohen et al. 1968; Chetverikov et al. 1976). It was shown that the intensity ratio for the main bands of blue luminescence is very dependent on the rate and direction of redox processes. The parameter $\xi = \frac{I_{550-0.5}I_{465}}{I_{465}}$ was proposed by Karnaukhov (1978) which allows one to estimate quantitatively the relationship between the oxidized and reduced forms of dehydrogenases. This parameter was used to solve some applied problems in the ecology of water systems (Karnaukhov et al. 1984) and in other cases (Chetverikov 1976; Chirkova et al. 1977).

We have shown that for plants growing on and out of deposits of heavy metals (Fig. 2), the leaf fluorescence spectra are different in the relationship between fluorescence bands in the blue-green region (Il'in et al. 1986). The relative intensities of luminescence are also different in the blue-green and red regions of the spectra of plant leaves grown under enhanced sulfate salinity conditions (Il'in and Kochubey 1987) and under conditions of water and nitrogen deficiency (Fig. 3). For the latter case, it is possible to select such pairs of bands over the whole range of the flourescence spectrum that, by changing the ratio of their intensities, it is possible to

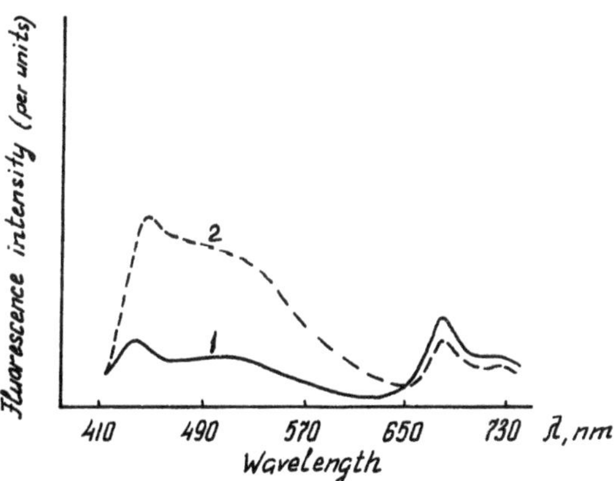

Fig. 2. The luminescence spectra of wormwood from the background site (*1*) and rare-earth metal deposit (*2*). *1* 1×10^{-3} %W, 5×10^{-4} %Mo, 5×10^{-4} %Sn; *2* 1×10^{-2} %W, 4×10^{-3} %Mo, 4×10^{-3} %Sn. (Il'in et al. 1986)

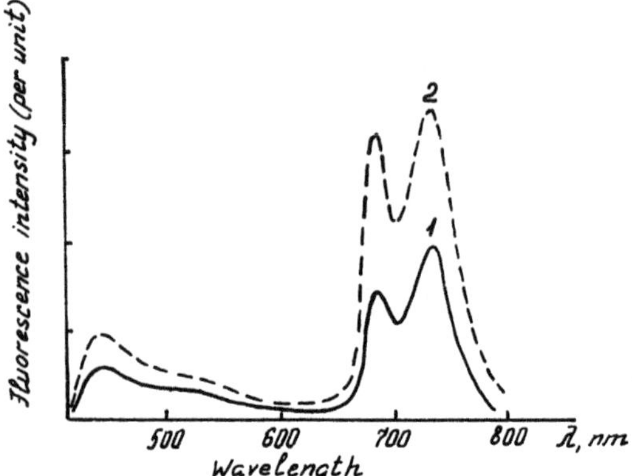

Fig. 3. The luminescence spectra of 2-week-old spring wheat plant leaves. *1* Control; *2* 0.1% sulfate in the soil. (Il'in et al. 1987)

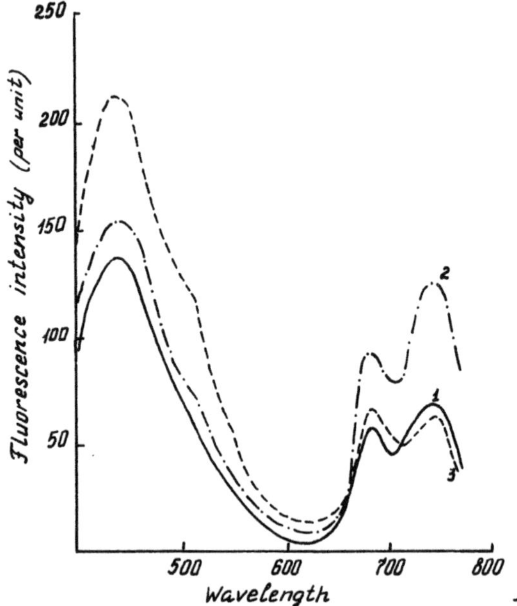

Fig. 4. The luminescence spectra of corn leaves. *1* Control; *2* water deficit; *3* nitrogen deficiency. (Shadchina and Beloivan 1992)

distinguish between different types of stresses (Fig. 4; Shadchina and Beloivan 1992). Thus, an analysis of the data in the literature shows that there are metabolites in plant cells emitting blue luminescence with a sufficiently high yield (Duysens and Amez 1957; Estabrook 1962; Siano and Muthazasan 1989). There are also publications on the application of the blue emission to examine cell and plant conditions (Chirkova et al. 1977; Karnaukhov 1978; Chappelle et al. 1984, 1985). However, there are some artifacts which can significantly decrease the efficiency of such testing, e.g., the emission of cellular walls which are not sensitive to metabolic processes, and reabsorption of some parts of the blue emission in the chlorophyll Soret band. Thus, higher levels of data distortion can be induced when the concentration of blue luminescence emitters is low. However, even in this case, their intensity can be as high as chlorophyll fluorescence because blue luminescence is not quenched by energy migration as for chlorophyll. An increase in luminescence intensity of compounds contained in cells with a low concentration can be caused by sensibilization, for example, by protein components. Reabsorption in the Soret band can be decreased by singularities of the spatial arrangement of chloroplasts, mitochondria, and other organelles containing blue luminescence emitters in cells and tissues. Thus, testing methods using blue luminescence must be developed further and laboratory investigations should continue. Measuring the luminescence spectra of various cellular organelles will provide the possibility to evaluate cellular wall emission. Stimulation or inhibition of organelle functions will help to distinguish luminescence from different metabolites and to evaluate the screening effect of chlorophyll absorption.

Stresses of various types such as salinization, the effect of heavy metals and drought, and the deficiency in a number of mineral elements necessary for plant nutrition such as nitrogen, phosphorus, and iron decrease the chlorophyll content in plant leaves (Sud'ina 1964; El-Sharkawy et al. 1986; Öquist 1986; Schmid and Feucht 1986; Canto de Loura et al. 1987). This is due to the decreased nitrogen content in the plant tissue caused by the reduced activity of nitrate reductase (Billard and Boucard 1982; Safaraliev et al. 1984). This results in the reduced biosynthesis of chlorophyll. We have shown that under favorable growth conditions wheat leaves exhibit some upper limit in chlorophyll content at certain stages of development. This also seems to be true for other plants. Therefore, by observing a decrease in pigment content in leaves as compared with its content under favorable conditions, it is possible to examine the stress. These data can be obtained by laboratory or remote measurement (McFarlane et al. 1980) of plant flourescence in the red region. When remote measurements are made, it is necessary to employ other data, including blue luminescence response, to determine the type of stress.

2.2 Laser-Induced Airborne Fluorometers

Remote measurements of fluorescence spectral characteristics of plants are carried out with the aid of specially designed airborne systems (Kim 1973; Kondratyev et al. 1987). Their major elements are lasers, used to induce fluorescence emissions from plants, and spectrometers used for remote spectral analysis of laser-induced fluorescence (Fig. 5). Taking into account the specific character of the laser-induced fluorescence emission spectra, the low quantum efficiency of fluorescence and the special features of remote sensing from airborne platforms, it becomes clear that the most suitable variant for remote sensing is a high-power, pulsed laser operating in ultraviolet and visible-light spectra and having sensing pulses with a repetition rate in the order of tens of pps. The most widely used lasers for this purpose are the 308-nm XeCl and the 355-nm Xe-F excimer lasers, the 266-, 532- and 255-nm YAG lasers, the 337-nm N_2 laser, and the dye lasers operating in the near-ultraviolet (350 nm) range to the red spectral regions (700 nm).

The major elements of remote spectrometers, used to measure and analyze the laser-induced fluorescence spectra, are: an optical receiving system and a storage-type spectrum analyzer with a data preprocessor (Fig. 5). A typical optical receiving

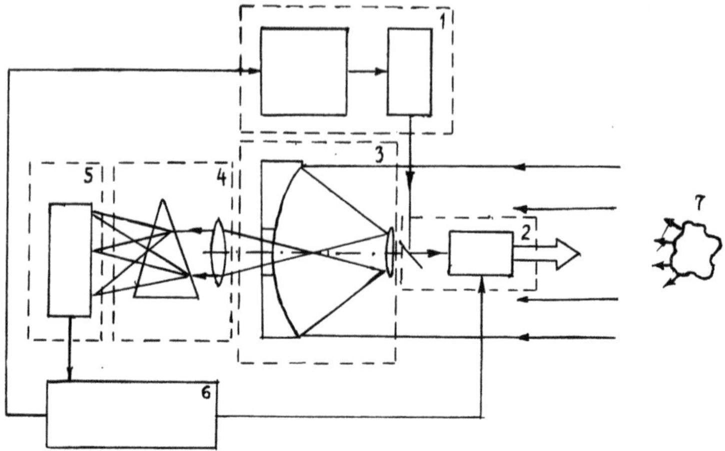

Fig. 5. Block diagram of an airborne laser fluorometer. *1* Laser with power unit; *2* collimator; *3* Cassegrainian telescope; *4* polychromator; *5* a set of photodetectors; *6* preprocessor; *7* target

system consists of a Cassegrainian reflecting telescope with a receiver mirror several centimeters in diameter.

The spectrum analyzer consists of a dispersing element (Fig. 5) which projects the spectrally analyzed light flux of fluorescence emission on to a set of photodetectors (Fig. 5). Photomultiplier tubes, photodiodes, or charge-couple devices may be used as photodetectors to convert a light flux into an electrical signal. Analog output obtained from photodetectors may be converted into digital form and can be processed by on-board computers.

2.3 Utilization of Airborne Laser-Induced Fluorometers for Remote Sensing of Vegetation

The first experiments regarding laser-induced fluorescence (LIF) spectrometry from helicopters were conducted in the United States in 1973 (Kim 1973). Multipurpose use of airborne laser-induced spectrofluorometers was carried out in Canada (Bristow et al. 1973), West Germany (Gehlhaar et al. 1981), Australia (Penny et al. 1986), and the Soviet Union (Bunkin et al. 1987; Kondratyev et al. 1987).

Lake algae were chosen for the first field experiments using laser remote sensing (Kim 1973). Subsequently, the majority of works in this field was carried out with sea algae (Gehlhaar et al. 1981; Penny et al. 1986).

2.3.1 Remote Sensing of Terrestrial Vegetation

Studies on terrestrial vegetation were initiated in the USA (Frank et al. 1983) utilizing airborne oceanographic lidar. The flights were operated at an altitude of 150 m and with a speed of 100 m s^{-1}. A combination of two lasers was used for excitation of fluorescence emission from natural trees and grasses: a 3-MW peak power frequency-doubled Nd:YAG laser operating at 532-nm wavelength with a 6-pps pulse repetition rate and a 422-nm XeCl excimer pumped dye laser with a 100 kW/pulse maximum output power. Concurrently with the field studies a series of laboratory experiments were carried out using the same lasers to obtain the entire spectral waveforms of the typical vegetation species growing along the flight lines. On the basis of the laboratory investigations (Fig. 6) two of the most informative spectral ranges (685 and 730 nm) were chosen, corresponding to the two major fluorescence emission peaks in the examined plant spectra.

Figure 7 shows profiles of airborne 532-nm laser-induced fluorescence. The variations in the fluorescence signal intensity obtained at the two chosen wavelengths (685 and 730 nm) during flights over the plant-covered terrain are practically identical. When passing over the plant-free parts of the terrain, the intensity level of the fluorescence signals was significantly lower in the chosen spectral region. Figure 8 depicts similar results obtained for the 685-nm fluorescence profile data with 422-nm excitation using the dye laser.

The above experiments were purely methodological, no attempts were made to determine any quantitative relationships between the characteristics of fluorescence signals and plant canopies. However, these experiments allowed one to optimize the performance of the airborne laser fluorometer in order to obtain reliable signals from different types of terrestrial vegetation.

The first measurement of laser-induced fluorescence emission from terrestrial plants was conducted in the Soviet Union in 1983 using a helicopter (Kanevski et al. 1985). The flights were carried out with a Ka-26 helicopter at an altitude of 30 m and a speed of 50 m s^{-1}. A 441.6-nm HeCd laser with output power of 20 mW was used to excite fluorescence.

Flourescence emission was detected at two wavelengths: 685 and 730 nm. The flights were operated in the early morning hours when the sun was low. Ground-based measurements of a series of essential crop characteristics such as total chlorophyll "a" and "b" concentration in wheat leaves, phosphorus concentration in leaves, dry biomass per unit crop area, and plant density were conducted concurrently with the flight measurements (Kanevski et al. 1985).

Figures 9 and 10 show the quantitative dependencies of crop fluorescence characteristics on the ground-based measurements of crop parameters. The ratio of fluorescence intensities at the two wavelengths, 685 and 735 nm, was chosen as an informative parameters.

In 1984 the Institute of General Physics of the Academy of Sciences conducted flight experiments using the airborne laser fluorometer *Chaika* (sea gull) (Bunkin et al. 1987). The cotton crop was chosen as the object of investigation. Fluorescence emission from the crop was induced at the second harmonic (532 nm) by the 150-mJ pulse energy YAG laser at a 10-pps repetition rate. The detecting part of the fluorometer included an optical receiving system based on the Cassegrainian reflecting telescope (with a receiving aperture diameter of 280 mm) and the multichannel optical spectroanalyzer (OSA) with parallel accumulation of fluorescence signals in different spectral channels.

Figure 11 shows typical chlorophyll fluorescence spectra of cotton crops measured from the aircraft. The differences between fluorescence spectra (at 730–750 nm) of intact crops and defoliated crops are quite clear.

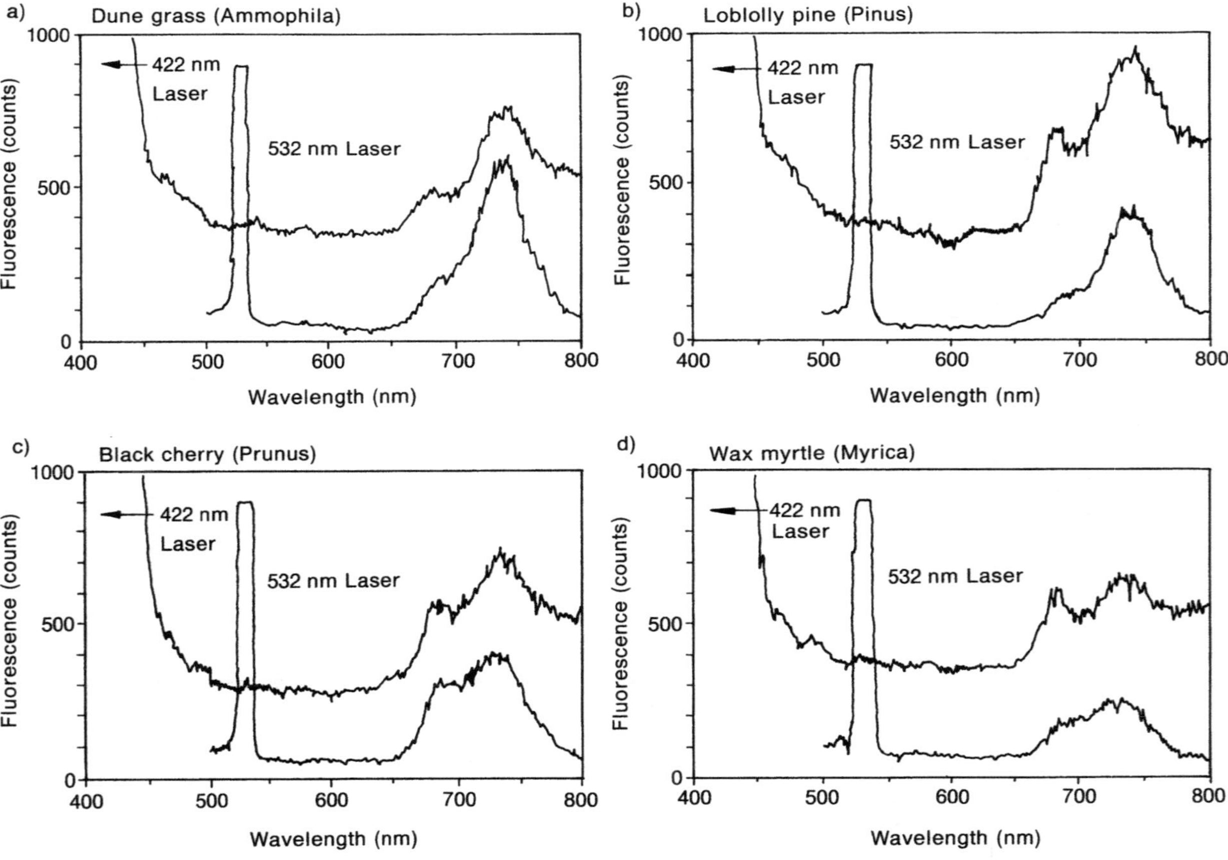

Fig. 6. Frequency-doubled Nd:YAG (532 nm) and 422-nm dye laser-induced fluorescence spectra of **a** dune grass, **b** loblolly pine needles, **c** black cherry, and **d** wax myrtle leaves. The blue excitation is a preferred wavelength since a broader and potentially more useful emission spectrum is produced. (Chappelle et al. 1985)

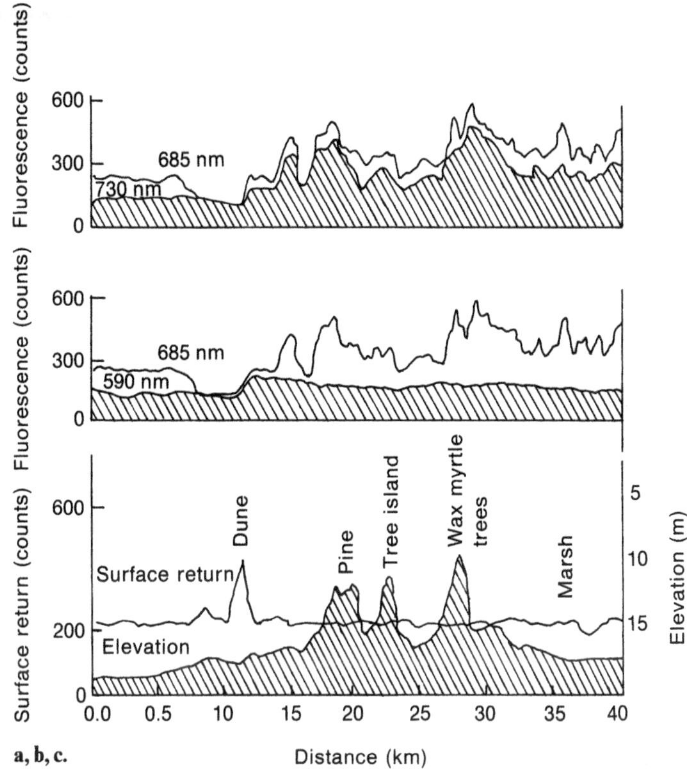

Fig. 7a–c. Profiles of airborne 532-nm laser-induced fluorescence at **a** 730 and 685 nm and **b** 590 and 685 nm, **c** terrain elevation and wavelength laser backscatter (surface return). (Frank et al. 1983)

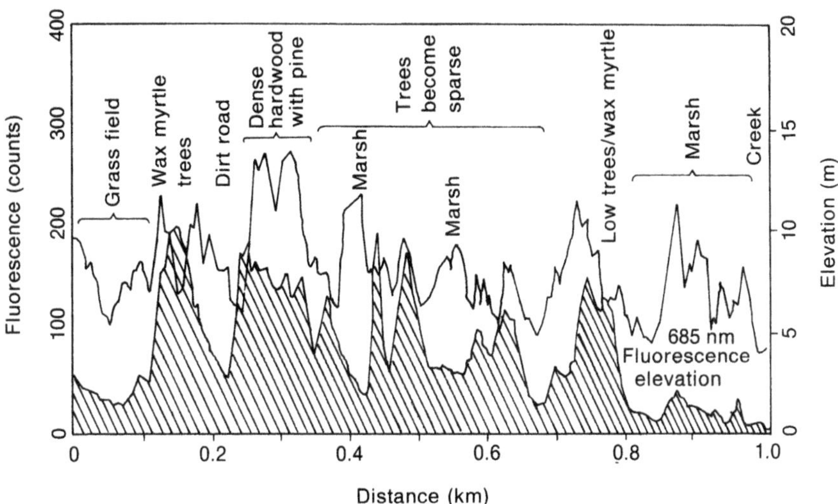

Fig. 8. Profiles of terrain elevation and 422-nm laser-induced fluorescence emission at 685 nm

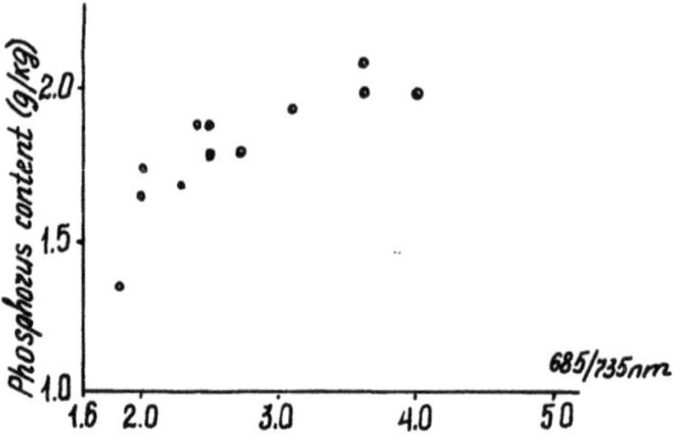

Fig. 9. The dependence of the intensity ratio of the winter wheat fluorescence at two wavelengths (685 and 735 nm) on the content of phosphorus in its leaves (g/kg dry weight)

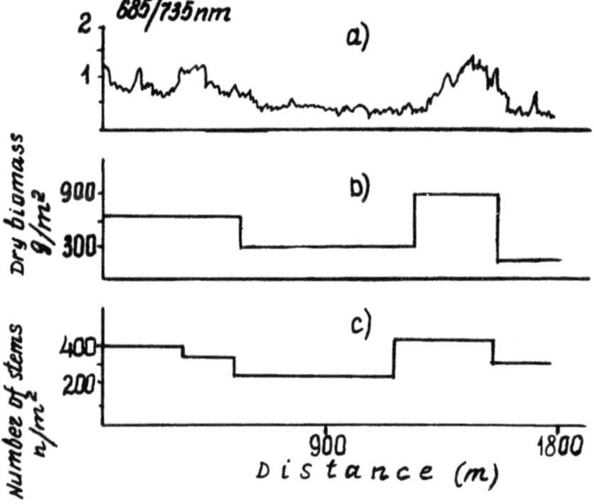

Fig. 10. Measurements of the fluorescence characteristics of the wheat crop depending on the density and biomass of the crop. **a** The variation in the 684- and 735-nm fluorescence intensity ratio. **b** The variation in biomass. **c** The variation in crop density

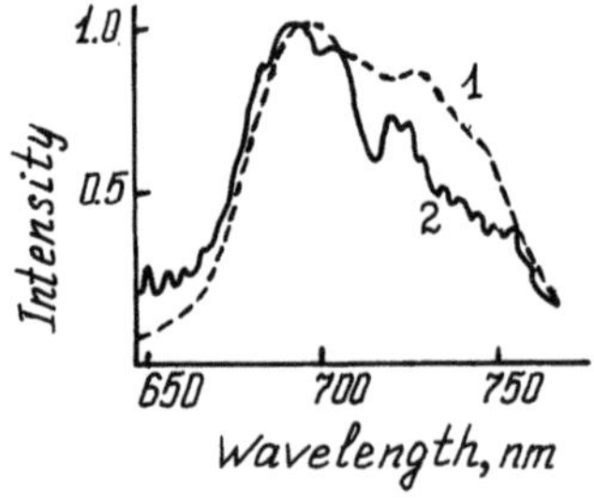

Fig. 11. The fluorescence spectra of *1* an intact and *2* defolliated cotton plant, obtained during the *Chaika* airborne fluorometer tests. (Bunkin et al. 1987)

In the USSR, the first data obtained by means of LIF emission from vegetation located on geological testing grounds were obtained in May 1985 (Kanevski et al. 1988). Two testing grounds were chosen for the experiment:

1. The molybdenum-silver mineralization was dated back to gneissic Proterozoic granites. The dominating vegetative canopy of the testing ground was Scotch pine. Mo-Ag background percentages were 3×10^{-4} and 6×10^{-6}, respectively, whereas within the confines of the deposit the percentages of molybdenum and silver increased to 2.2×10^{-3} and 4.5×10^{-5}, respectively.

2. The copper-molybdenum mineralization was dated from hydrothermally changed, eruptive, and intrusive Paleozoic rocks. The vegetation of the profiles investigated was mainly comprised of xerophytic, semidesert species dominated by wormwood. The content of metals in the diluvium in the background areas of the investigated profiles was 0.008 and 0.0007 for copper and molybdenum, respectively.

The flights were operated at an altitude of 50 m in the early morning hours. The flight speed of the Mi-8 helicopter was 100 km hr^{-1}. A 10-mJ pulse-energy N_2 laser with a 15-pps pulse repetition rate was used to induce fluorescence emission from the vegetation.

The results of measurements of the laser-induced fluorescence emission from vegetation covering the chosen testing grounds allowed an informative parameter to be determined which proved to be the ratio of two fluorescence peaks at 488- and 440-nm wavelengths. Figures 12 and 13 show variations in this parameter along the investigated profiles of the copper-molybdenum and molybdenum-silver mineralizations. As seen in Fig. 12, the ratio of fluorescence intensities at the chosen wavelengths for the background areas of the copper-molybdenum mineralization is 1.1 and has a natural tendency to increase up to 1.55–1.65 within the confines of the deposit. A similar tendency is found also for the molybdenum-silver mineralization where the ratio increases along the profile from 1.2 for the background areas to 1.8–2.8 for the areas with a high metal content in the diluvium (Fig. 13).

2.3.2 Remote Sensing of Aquatic Vegetation

The biological productivity of seas and oceans depends primarily on algal photosynthesis, which accumulates solar energy and transforms it into chemical bond energy of biological molecules. As a leading role in this process is taken by chlorophyll *a* molecules, one of the most important tasks of remote sensing of aquatic vegetation is

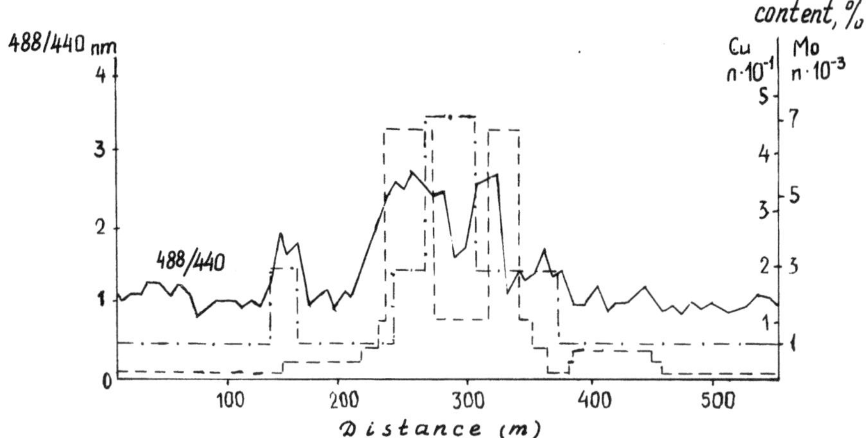

Fig. 12. The results of laser remote sensing of the copper–molybdenum mineralization from an airborne platform. The measured parameter was the ratio of fluorescence intensities 488/440 nm (per unit). —— Fluorescence intensity ratio, --- Cu percentages, –·–·– Mo percentages. (Kanevski et al. 1988)

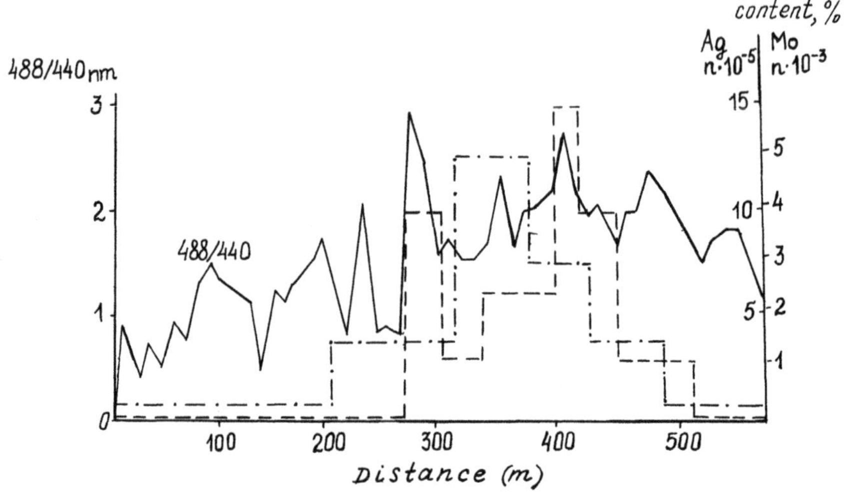

Fig. 13. The results of laser remote sensing of the silver–molybdenum mineralization from an airborne platform. The measured parameter was the ratio of fluorescence intensities 488/440 nm (per unit). —— Fluorescence intensity ratio, --- Ag percentages, –·–·– Mo percentages. (Kanevski et al. 1988)

the mapping of chlorophyll distribution in aquatic ecosystems. Fluorescence spectroscopy is an efficient technique of remote assessment of the chlorophyll content in the water.

The first detection of laser-induced fluorescence emission from aquatic systems was in the United States in 1973 (Kim 1973). A 590-nm, 0.25-J pulse-energy dye laser operating on board a helicopter was used during the experiment. The minimal chlorophyll *a* content in the water, the fluorescence emission from which was reliably detected from an altitude of 30 m, was a few fractions of a milligram per cubic meter. During the experiment, the chlorophyll *a* concentration in Lake Ontario, above which the flights were conducted, varied from 4 to 12 $mg\,m^{-3}$.

In the past 15 years a great number of investigations have been carried out in the field of laser remote sensing of fluorescence emission from sea algae from aircraft and sea vessels (Kim 1973; Bristow et al. 1973; Gehlhaar et al. 1981; Penny et al. 1986). However, the problem of quantitatively estimating the physicochemical and biological parameters of water through its fluorescence characteristics still remains to a great extent unsolved. By examining the typical spectra of laser-induced seawater echo signals (Fig. 14), one may observe the presence of several specific peaks. The first one (of the shorter wavelength) is associated with Rayleigh and Mie scattering and coincides with the laser line (in this case, 514.5 nm). The second 580-nm peak is associated with fluorescence of the algal pigment phycoerythrin. The third 625-nm peak coincides with a water Raman line displaced 3450 cm^{-1} to the right of the laser line. And the fourth 685-nm peak corresponds to the chlorophyll *a* fluorescence band.

It should also be noted that a vaguely expressed spectral peak at 530–540 nm reveals asymmetric, broad-band fluorescence from dissolved organic matter associated with both the decomposition of the algal biomass and organic matter. Figure 15

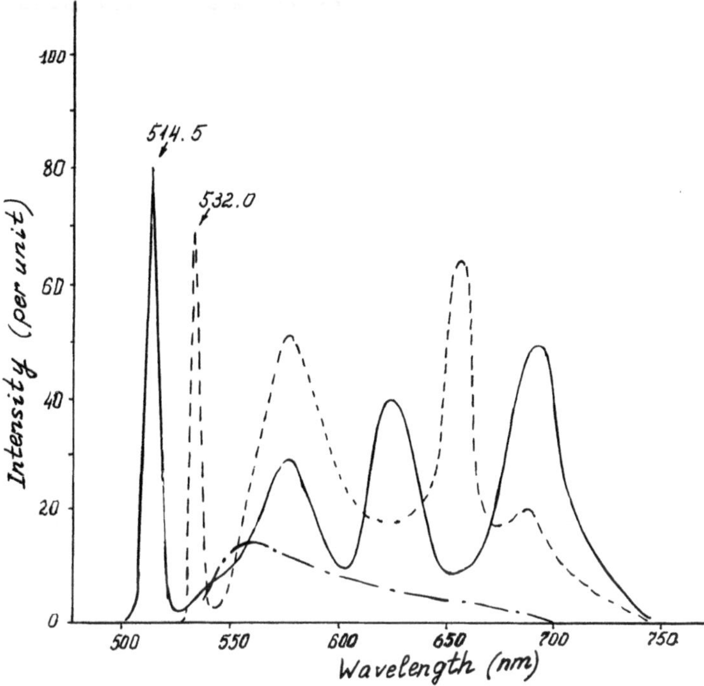

Fig. 14. The spectra of laser-induced emission from an aquatic surface at 514.5-nm (——) and 532.0-nm (– – –) wavelengths detected by the remote fluorometer. The fluorescence spectrum (–·–·–) attributed to the dissolved organic matter in the water

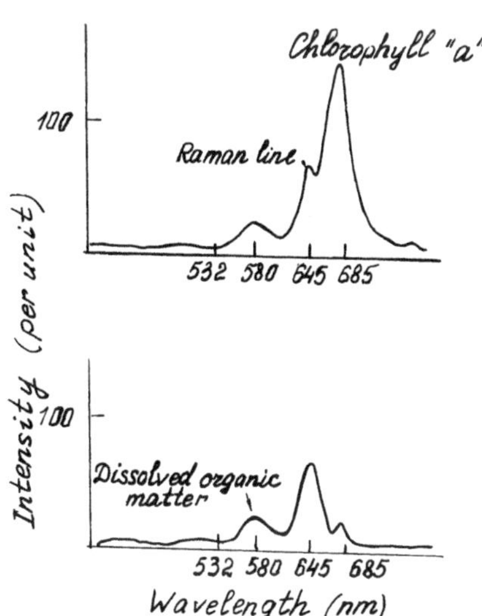

Fig. 15. Characteristic levels of the major spectral components of the echo signal detected by AOL project experiments (USA) (North Sea, 1978). (Bunkin et al. 1987)

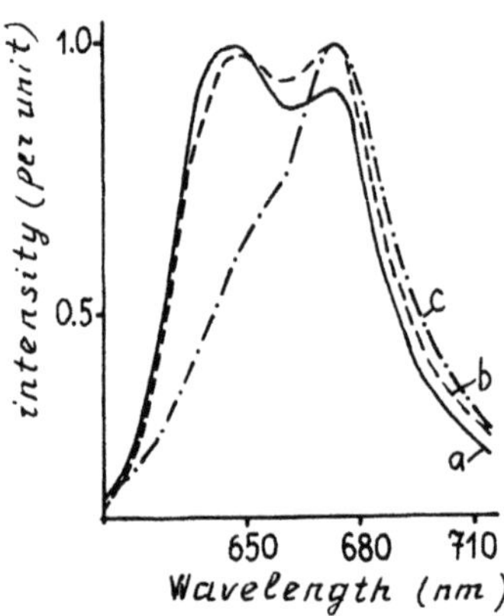

Fig. 16. The dependence of the spectral shape of phytoplankton *Synechococcus* fluorescence on the exposure to $HgCl_2$ addition: *a* control sample; *b* 1.5 min after the addition of $HgCl_2$; *c* 30 min after the addition of $HgCl_2$

shows typical spectra of echo signals detected from aircraft in the North Sea during flight tests of airborne oceanographic lidar (AOL) (Bunkin et al. 1987).

However, the attempts to develop a technique of remote assessment of the chlorophyll *a* concentration in the water by measuring the laser-induced, 685-nm chlorophyll fluorescence peak from air-craft or sea vessels were only successful in some particular cases. The major difficulty in assessing the chlorophyll concentration according to the measured value of the chlorophyll fluorescence intensity lies in the necessity of determining seawater attenuation factors for the wavelength of the laser beam itself and for the wavelength at which the LIF emission is being detected. However, for a number of cases the measure of the optical attenuation may be an indication of Raman scattering of laser-excited water molecules. The Raman scattering signal depends on the physicochemical properties of the water itself and is determined mainly by laser emission and fluorescence signal attenuation factors. Thus, if the value of the fluorescence signal is divided by the value of Raman scattering, there is the possibility of a strong correlation between the chlorophyll *a* concentration in the water and the ratio of chlorophyll fluorescence signal intensity. Therefore, the chlorophyll *a* signal should be normalized on the basis of Raman scattering for each pulse of laser emission.

The results of laser fluorescence studies of the blue-green algae exposed to mercurial contamination have a special value when applied to ecological problems. Thus, the addition of 0.01% solution $HgCl_2$ significantly disrupted alga photosynthesis and was easily detected on the basis of the change in the fluorescence spectrum induced by the 337-nm nitrogen laser (Fig. 16).

These results indicate potential future applications of remote fluorescence diagnosis of algae for ecological monitoring of aquatic ecosystems.

3 Laser Remote Sensing of the Geometrical Structure of the Plant Canopy

Laser beam interaction with plants results in absorption of part of the radiation energy by plants, while the rest is reflected. The phenomenon of fluorescence described in the preceding section is related to the radiation energy absorbed by plants. In this section the mechanism of laser beam energy reflected from plants is considered with the information provided by the characteristics of the scattering of radiation energy by the vegetation.

The study of the reflecting properties of the plant canopy over various spectral bands remains the key problem of conventional, passive techniques of plant remote sensing (Kondratyev and Fedchenko 1982). The major characteristic being measured in this case is the spectral reflectance of vegetation . This characteristic of vegetation is extremely variable and depends on a number of factors such as radiation regime, physicochemical properties of the plant environment, and the biological properties of the plants (Kondratyev and Fedchenko 1982). Therefore, the assessment of vegetation conditions by means of reflective characteristics is difficult.

3.1 Bidirectional Reflectance Factor and the Geometrical Structure of the Plant Canopy. Theoretical Calculations

The first conclusions to emerge from conventional procedures of measuring the reflective characteristics of vegetation were based on measurements of the angular distribution of intensity of the solar radiation reflected from the vegetation (bidirectional reflectance factor, BRF) over certain spectral bands.

Mathematical models determine a number of the important geometrical parameters of plants by means of intensity measurements of the radiation reflected by plants in several fixed directions of scattering (Gerstl et al. 1986; Ross and Marshak 1988). The theoretical calculations of BRF for different types of plant canopies (Kanevski and Ross 1983; Ross and Marshak 1988) enable the most informative scattering directions to be determined in order to obtain such geometrical parameters as leaf area index, crop height, leaf sizes, etc. Later, this approach stimulated the emergence of a basic new trend in remote sensing techniques, i.e., the laser sensing of plant canopy architecture.

Let us consider the major results of BRF calculations for different vegetation types. Figures 17 and 18 depict the BRFs of a spruce fir crown having various geometric shapes and of the monolayer (Kanevski and Ross 1983). The spruce crown was modeled as a multilayered cone. Each layer of the cone was filled with diffuse reflecting cylinders imitating spruce branches. The leaf monolayer was modeled by diffuse reflecting cylinders with their centers lying in the same plane, whereas the axes were obtained randomly. The calculations confirmed the basic distinction between the leaf monolayer and the spruce crown BRF. The greatest differences in the BRFs calculated for various types of model structures became especially apparent in the narrow range of backscattering angles. In the literature this angle range is usually called "hot spot" (Gerstl et al. 1986). The results of modeling the crop BRFs are of special interest. This modeling approach allowed one to study the influence of various geometrical structures of the crop on the BRFs and to evaluate the most informative scattering directions with respect to such important crop characteristics as height, effective distance between the leaves, leaf sizes, and individual properties of the geometrical plant structure (Ross and Marshak 1988; Kondratyev et al. 1987).

The plant canopy for which the calculations were performed was modeled by a set of plants arranged in a checkered manner. It was proposed that their round,

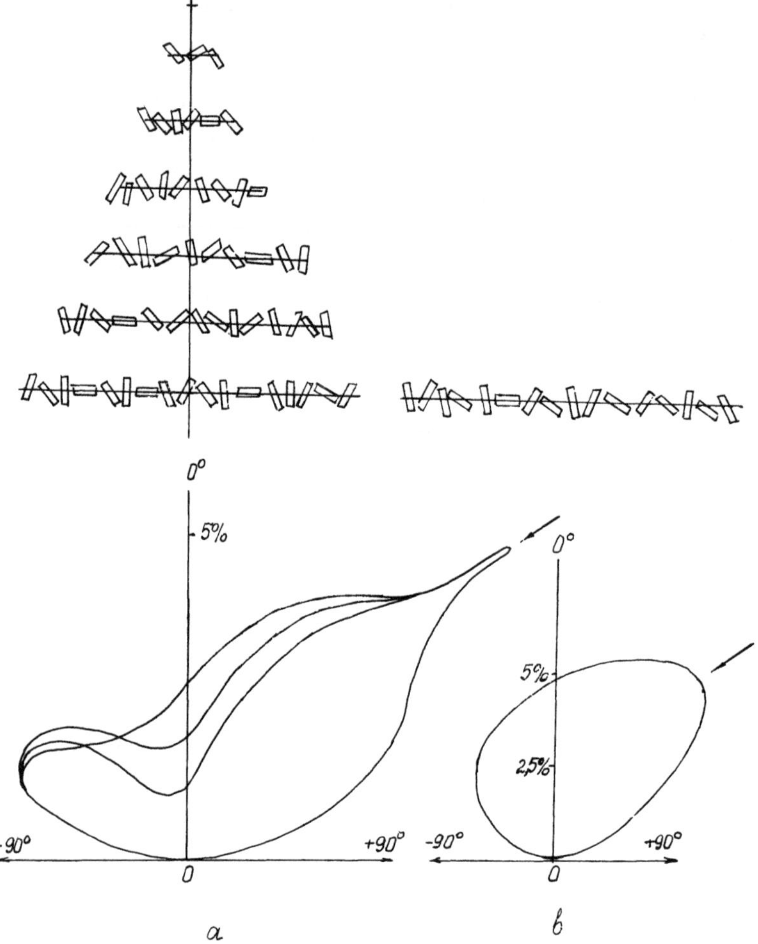

Fig. 17. The BRF of the spruce fir crown (**a**) and the monolayer from spruce branches (**b**). The direction of the solar beams is indicated by *arrows*. The *upper part* of the figure expresses the model of the crown structure

diffuse reflecting leaves were horizontally located on the vertical, opaque stems. The major parameters of the model were as follows: spectral band of the scattered radiation of the plants, 380–710 nm; spectral reflectance of leaves and stems, $R = 0.1$; nonreflecting soil; distance between the plants, $A = 0.1$ m; crop density, 100 plants/m^2; crop height, $H = 1$ m; distance between the neighboring leaves on the stem, $Z_r = 0.1$ m; average number of leaves per plant, $N_L = 10$; angle of leaf rotation on the genetic spiral, $\alpha_r = 120°$; leaf diameter, $d_L = 60$ mm; stem diameter, $d_S = 0.2$ mm; leaf area index of the crop, $L_L = 2.69$; stem area index of the crop, $L_S = 0.02$; spectral reflectance factor, b(v); zenith sighting angle, V. The results of canopy BRF calculations in the principal plane at various values of the plant canopy architecture allow the effect of crop height and the distance between the neighboring leaves on crop BRF to be understood. Figure 19 shows a dependence of the BRF of

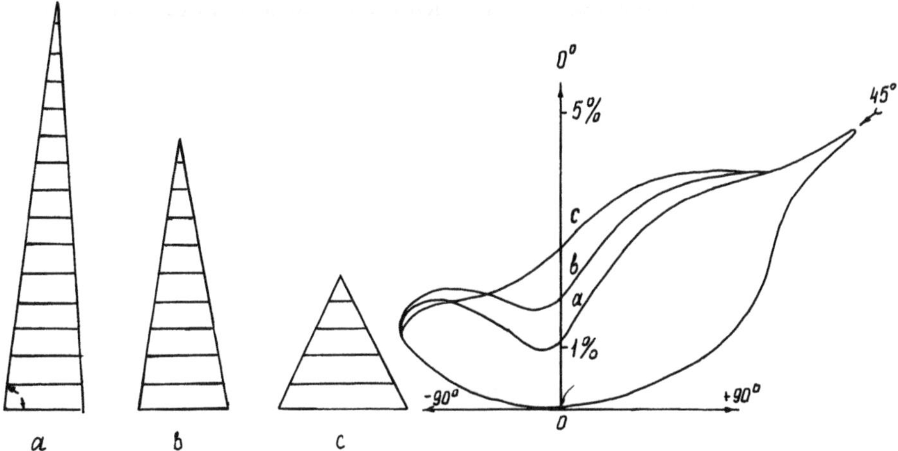

Fig. 18. The BRF of the three types of spruce fir crowns

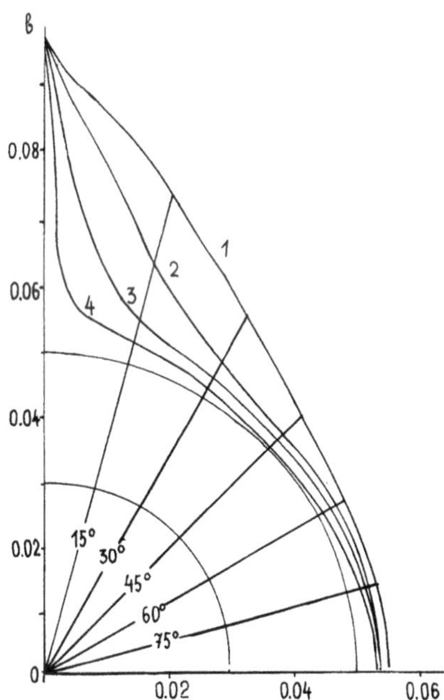

Fig. 19. The dependence of the crop spectral reflectance factor, b (v), on the zenith sighting angle, V, at different crop heights. Solar zenith angle, $V_0 = 0$; genetic spiral angle, $\alpha_r = 60°$, $N_r = 12$, $L_1 = 3.25$. The other parameters are standard. Curves are denoted *1*, *2*, *3*, and *4*. The distances between the leaves of the stem, Z_r, are 2.5, 5, 10, and 40 cm. The crop heights, H, are 30, 60, 120, and 480 cm

the crop on the view zenith angle at $\varphi = \varphi_0$ and $\varphi = \varphi_0 + 180°$ and at the sun zenith angle, $V_0 = 0$. The calculations are given for $\alpha_r = 60°$. In the case under consideration the leaves of each plant overlap each other as well the leaves of the neighboring plants. The first leaf overlaps the seventh one, the second leaf overlaps the eighth one, and so on. The calculations were made for the cases when the distance between the leaves varied from 2.5 (Fig. 19, curve 1) to 40 cm (Fig. 19, curve 4); the crop height varied from 30 to 480 cm, respectively.

It is quite clear from Fig. 19 that when the leaf area index (LAI) is constant, the decrease in crop height results in an increase in its BRF, especially in the hot spot area.

Crop height also influences the width of the hot spot area. In the case of high crops the hot spot is very narrow and the crop may be considered as a turbid medium.

In the case of low crops the difference between the BRF of the model crop and the turbid medium BRF is rather significant and the width of the hot spot area spreads over 60° from the nadir. The 0.05 radius circle in Fig. 19 corresponds to BRF of the turbid medium layer.

Leaf Dimension Effect. The finite leaf dimension is one of the major differences between the model crop and the turbid medium. A number of experiments conducted to estimate the effect of diameter d_L of the round leaf on the BRF showed that an increase in the leaf diameter d_L was inevitably followed by a decrease in the distance A between the plants, whereas the leaf area index and the crop height remain constant. Consequently, in the case of small leaves, when the crop density is high, the structural element role of the plant decreases and the crop tends to resemble a microdispersion turbid medium.

In the case of big leaves, the crop density is small, but the structural role of the plant also decreases and the crop tends to resemble a macrodispersion turbid medium. It follows from Fig. 21, representing the results of the calculations, that the increase in the leaf diameter causes the increase in the crop brightness along with an expansion in the hot spot area. Outside the hot spot area the BRF resembles the BRF of the diffuse reflecting surface.

A comparison of Figs. 20 and 21 shows that an increase in either the leaf diameter or the distance between the neighboring leaves may equally enlarge the hot spot area. This is attributed to the fact that when the leaves are mutually shaded, it is their angular dimensions that count regardless of whether their increase is caused by enlargement of the leaf diameter or by the decrease in the distance between the leaves.

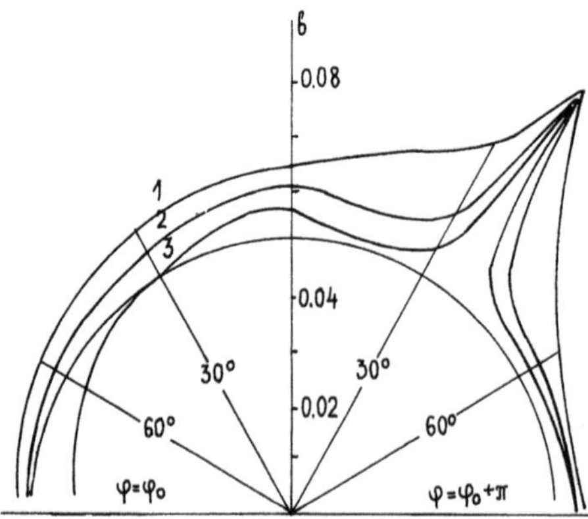

Fig. 20. The dependence of crop, b(v), on the zenith sighting angle, V, at different crop heights. The parameters are standard. Curves are denoted *1*, *2*, and *3*. The distances between the leaves on stem, Z_r, are 2.5, 5, and 20 cm. The crop heights, H, are 25, 50, and 200 cm

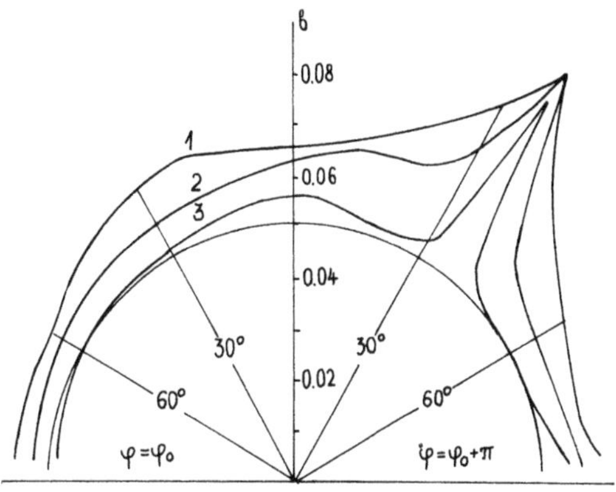

Fig. 21. The dependence of crop, b(v), on the zenith sighting angle, V, at different leaf diameters. Curves are denoted *1*, *2*, and *3*. The leaf diameters, d_L are 16, 4, and 2 cm. The distances between the plants in the row, A, are 27, 7, and 3 cm

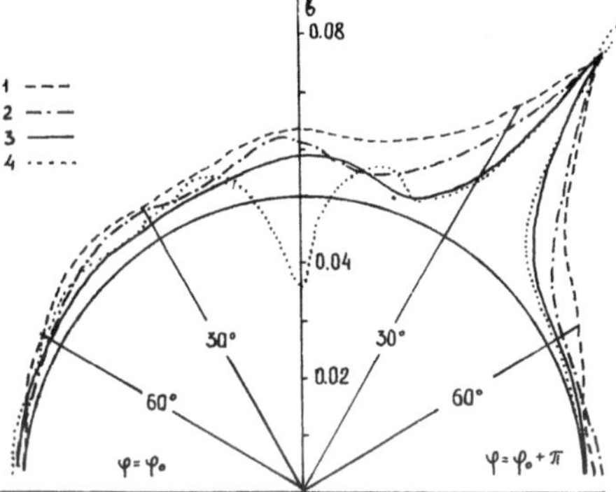

Fig. 22. The dependence of crop, b(v), on the zenith sighting angle, V, at different number of crop layers. Curves are denoted *1*, *2*, *3*, and *4*. The leaf diameters, d_L, are 12.5, 9, 6, and 4 cm. The distances between the leaves on stem, Z_r, are 40, 20, 10, and 5 cm. The number of layers in the crop: 2.5, 5, 10, and 20

Crop Structure Effect. In order to assess the effect of the plant structure on the BRF at a constant crop height and LAI, the leaf diameter and, accordingly, the distance between the neighboring leaves on the stem were decreased. When the leaf diameter is small, the effective plant diameter is also small; the leaves are located close to each other and the plant itself is comparable to a slim cylinder tightly packed with small leaves. The distance between the plant rows is relatively large. When a separate plant is dominant, as in this case, the crop may hardly be compared with a turbid medium. And vice versa, if the leaf diameter is large, the plant has only a few large leaves, the role of a separate plant in the crop architecture is insignificant, and the crop consists of several layers of large leaves.

Thus, the extreme values of the leaf diameter correspond to the extreme variations in the crop architecture. The effect of the above-mentioned dependencies on the BRF is shown in Fig. 22. The hot spot effect is the most strongly pronounced in the case of a multilayered structure of the crop (see curve 1) and is hardly pronounced in the case of a cylindrical plant structure (see curve 4). The wide and deep minimum of the BRF in the nadir for the cylindrical plant structure is caused largely by the dark soil effect. The secondary minimum at $V = 20°$, $\varphi = \varphi_0 + 180°$

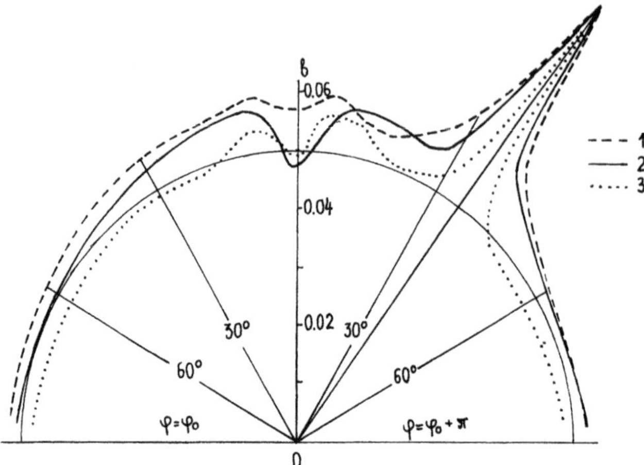

Fig. 23. The dependence of crop, b(v), on the zenith sighting angle, V, at different ellipticities of the leaf. The soil brightness, b = 0.1, other parameters are standard. *1* Round leaf, $d_{L1} = d_{L2} = 3$ cm; *2* moderately elliptical leaf, $d_{L1} = 6$ cm, $d_{L2} = 1.5$ cm; *3* strongly pronounced ellipticity of the leaf, $d_{L1} = 9$ cm, $d_{L2} = 1$ cm

should be apparently attributed to a particular value of the solar azimuth $\varphi_0 = 30°$, the mutual shading of the plant cylinders being at its maximum.

The Leaf Shape Effect. The large majority of numerical experiments deal with circular leaves. However, circular leaves are not common to monocotyledonous crops, such as the cereals. Hence, it was interesting to investigate the effect of the leaf shape on the crop BRF. For this purpose three variant calculations were performed: (1) for circular leaves 0.03 m in diameter; (2) for moderately circular leaves with semi-axis lengths of 0.06 and 0.015 m; and (3) for elongated elliptical leaves with semi-axis lengths of 0.09 and 0.01 m. The analysis of the calculations shown in Fig. 23 indicates that as the elliptical leaf shape becomes more and more pronounced, the crop brightness starts to decrease, especially in the hot spot area. The hot spot area also decreases, while the zenith minimum of the BRF is due to the fact that, according to the calculations $b_n = 0.1$.e., the soil is optically black. On the basis of Fig. 23, the general conclusion can be made that the turbid medium model provides a much better approximation of a crop with long narrow leaves than that of a crop with circular leaves.

The Angle of Leaf Rotation on the Genetic Spiral. The angle of rotation on the genetic spiral, α_r, is an important shape-making parameter of the plant which sets a pattern of leaf arrangement on the stem. It is this parameter that determines the possibility to sight the lowermost leaves or soil in the given direction. For example, let us consider a purely theoretical case, nonexistent in nature, in which all leaves are arranged on the same side of the stem exactly one above the other. In this case only the uppermost leaf could be sighted in the direction of the nadir, the mutual shading of the leaves being at its maximum. If $\alpha_r = 180°$, all the odd leaves would be on one side and all the even leaves on the opposite side of the stem, in which case the two uppermost leaves could be sighted in the direction of the nadir, the rest of the leaves being right underneath them.

Circular leaves do not shade each other in the direction of the nadir, when $\alpha_r = 360°$ and $180°$, within one spiral coil. However, with the further diminution of α_r the mutual shading of the leaves starts within one coil. The values of the angle of rotation, α_r, of the distance between the leaves, Z_r, determine the thickness of the active upper layer of the crop, $Z_B = 360\, Z_r/\alpha_r$, where the mutual shading of the leaves is minimal. The leaves, which are below the active layer in the direction of the nadir, have not been sighted as yet and hence cannot take part in forming the spectral reflectance factor.

Figure 24 depicts a cross-section of the crop BRF in the principal plane. Crop brightness is minimal at $\alpha_r = 360°$, when only the uppermost leaf of a plant is illuminated on the black soil background, the rest of the leaves being shaded. The crop BRF corresponds to the diffuse reflecting surface BRF, i.e., the uppermost leaf BRF. At $\alpha_r = 180°$ only the two uppermost leaves are illuminated and the b(V) depends on the degree of shading of the lower leaves by the uppermost ones. The hot spot, observed in the near-zenith directions is not pronounced.

The further diminution of α_r results in increased brightness caused by the increase in leaf number, within one genetic spiral coil, each leaf making its own contribution to crop brightness. At the same time the probability to sight the non-reflecting soil diminishes. It becomes more complicated at other directions of sighting. The diminution of the angle of rotation on the genetic spiral, α_r, not only increases the number of the illuminated leaves but at the same time it increases the depth in location of the lowermost, illuminated leaves. Hence, the probability to sight these leaves is very low since they are shaded by the uppermost leaves. As a result of the interaction of the two counteracting factors, i.e., the quantitative increase in the illumination of the leaves, on the one hand, and the decrease in the probability to sight the lower illuminated leaves, on the other hand, the lateral brightness begins to increase (provided that α_r is decreasing), reaching its maximum at around $\alpha_r = 90°$ and then begins to gradually decrease.

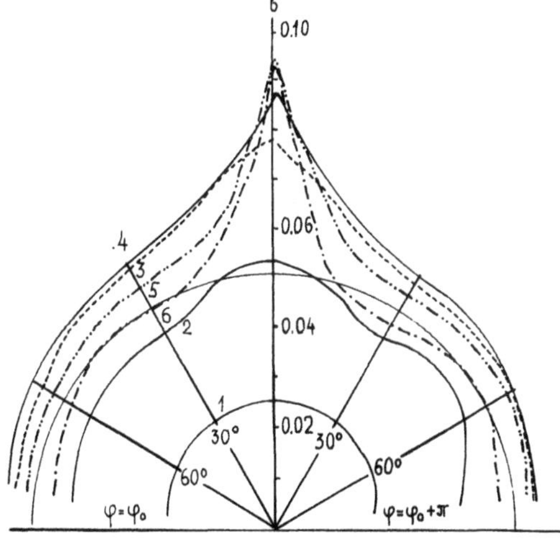

Fig. 24. The dependence of crop, b(v), on the zenith sighting angle, V, at different genetic spiral angles of rotation, α_r. Sun zenith angle, $V_0 = 0°$, other parameters are standard. *1* $\alpha_r = 360°$; *2* $\alpha_r = 180°$; *3* $\alpha_r = 120°$; *4* $\alpha_r = 90°$; *5* $\alpha_r = 45°$; *6* $\alpha_r = 60°$

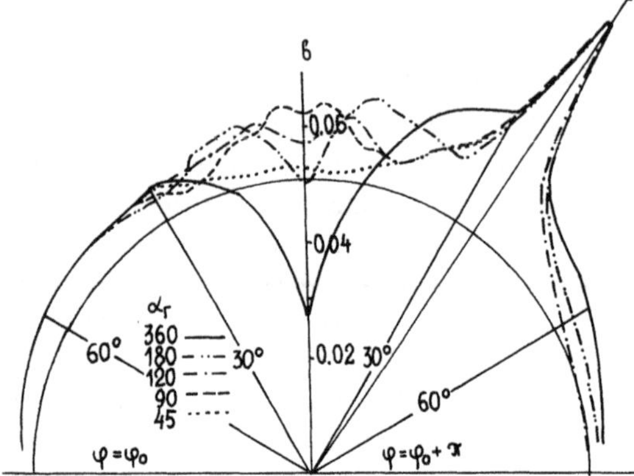

Fig. 25. The dependence of crop, b(v), on the zenith sighting angle, V, at different values of the genetic spiral

Figure 25 shows a cross-section of the BRF in the main plane at $V_0 = 36°$. The hot spot is strongly pronounced, but it is obvious that the BRF is unaffected by the angle of leaf rotation, α_r. The effect of α_r on the BRF is maximal in the nadir direction. The minimal brightness is reached at $\alpha_r = 360°$, caused by the fact that only the uppermost leaves can be seen while the remaining leaves are in shade, although some of them are illuminated. At further diminution of α_r the crop brightness is once more affected by the two counteracting factors. Due to the increase in leaf number in one genetic spiral coil, the probability to sight the lower leaves in the nadir increases. At the same time, the probability of their being shaded by these upper leaves of the neighboring plants, which are located in the direction of the solar path, increases. The domination of the first factor has its effect until the spectral brightness coefficient reaches its maximum at $\alpha_r = 90°$. The further diminution of α_r causes the domination of the second factor at which the spectral brightness coefficient decreases. A comparison of Figs. 24 and 25 provides the basis for the conclusion that the BRF in near solar directions largely depends on the sun zenith angle, V_0. The BRF is the most sensitive to variation in the angle of leaf rotation on the genetic spiral α_r in the direction of the nadir, but this dependence lacks uniqueness at different solar zenith angles, V_0.

The Effect of Stem on the Spectral Brightness Coefficient of the Crop. To evaluate the effect of the stems on the BRF the following numerical experiment was conducted. For the given crop parameters the effective stem area was increased and the leaf area correspondingly decreased, so that their total area remained unchanged. Curve 1 in Fig. 26 corresponds to the case when the stems are not taken into account. If the stem area is increased, the left side of the BRF starts to collapse. This phenomenon is caused by the fact that the vertical stems are opaque and so do not contribute to the spectral brightness coefficient at the azimuth, $\varphi = \varphi_0$. If the whole crop consists only of the stems, the brightness will be exclusively due to the upper end of the cylinder. The other effect caused by enlargement of the stem area is the decrease in quantity and width of the hot spot.

Finally, we present the results of calculations of the BRF corresponding to the various stages of the cereal crop modeled by plants with nonhorizontal, elliptical leaves with an inclination angle of 60°, 0.16 m long, and 0.01 m wide. The distance

Fig. 26. The dependence of crop, b(v), on the zenith sighting angle, V, at different proportions of the area index of the vertical stems in the total crop LAI. Curves are denoted *1*, *2*, *3*, and *4*. Stem area indices, L_S, are 0, 1, 2, and 2.71. Leaf area indices, L_L, are 2.71, 1.71, 0.71, and 0. Leaf diameters, d_L, are 3, 2.4, 1.5, and 0 cm

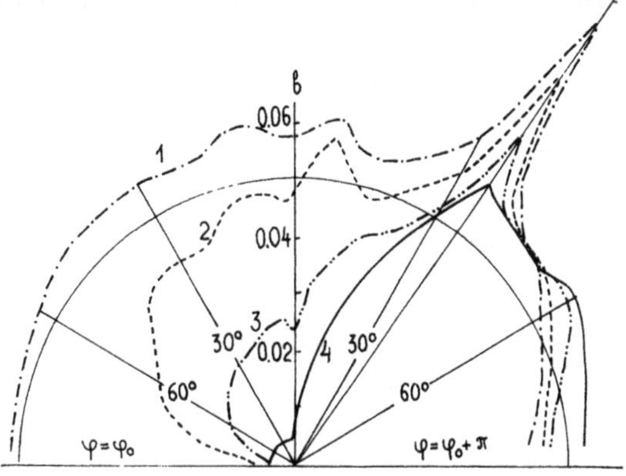

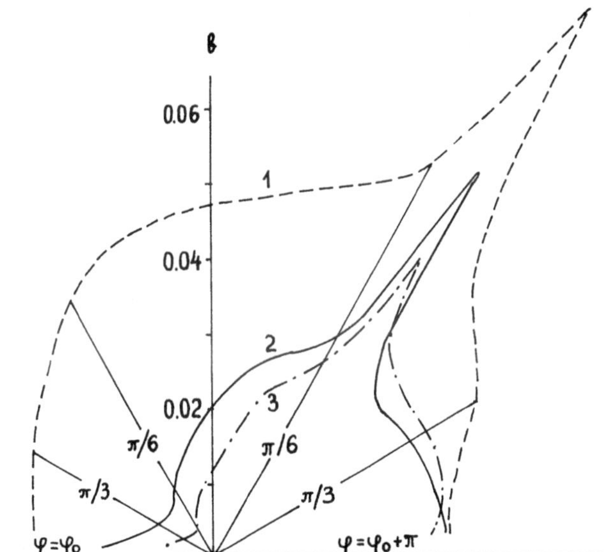

Fig. 27. Polar diagram of the canopy BRF on the principal plane at different growth stages. Solar zenith angle, $\theta_0 = \pi/5$; dry soil; red spectral interval, soil reflectance $b = 0.14$

No. of curve	Crop stage	Crop height (cm)	Leaf area index	Stem area index
1	Tillering	20	0.875	0
2	Stem elongation	40	2	0
3	Heading	80	4	1.9

between the plants is 0.10 m; the crop density is 400 plants/m^2; the angle of leaf rotation on the genetic spiral, α_r, is 180°; the stem diameter is 0.0006 m; the height of location of the first leaf from the ground is 0.089 m. As can be seen in Fig. 27 the well-pronounced BRF asymmetry effect is typical of the crop with nonhorizontal leaves.

Various stages of plant growth cause some significant changes in the hot spot area of the BRF.

A number of numerical experiments that have been conducted with crops allow the sensitivity of the crop BRF to the variations of its geometrical parameters to be investigated.

The dependence of the BRF on the geometrical structure of the plant canopy holds true for both the model crop and the tree crown. This dependence is especially strongly revealed in the backscattering region. Consequently, one of the major conclusions of great value regarding practical application is the confirmation that the measurements of the BRF plant canopy parameters in the hot spot area can be used to obtain data concerning the vegetation architecture. It is the region of the BRF which demonstrates the discrete character of the plant structure in the most complete way and that is the most sensitive to leaf spatial location, which is closely related to such important plant canopy parameters as the growth stage, plant, biomass, and leaf area index. Therefore, the practical implementation of the results of theoretical investigations and the development of techniques for the remote measurement of the BRF backscattering region become the most acute tasks in the study of remote sensing techniques.

We will now consider the experimental investigations that have already been conducted in this area.

3.2 Laser Measurement Techniques of Plant BRF in the Hot Spot Region

The above-mentioned results of mathematical simulation justify the use of remote measurements of the BRF structure in the hot spot angular regions (1°–10°) for obtaining data on the geometrical parameters of the plant canopy. The measurement of the BRF structure in its hot spot area from aircraft by conventional passive techniques is fraught with technical difficulties which make it practically impossible. Their enumeration is given below. When the flight is operated at low altitudes (up to 100 m), the angular dimensions of the shadow cast by the aircraft are comparable with the hot spot region, thus impeding the measurements of the BRF hot spot area (Fig. 28). Flights at altitudes greater than several hundred meters using scanning spectrometers with a narrow view angle avoid the above difficulties, but a series of new problems arise caused by the need to keep the aircraft perpendicularly aligned to the principal plane. This is necessary in order to obtain a cross-section of the BRF exactly in the backscattering region. Additionally, the use of narrow angle scanning spectrometers at great altitudes substantially complicates the interpretation of the data obtained because a change in the viewing angle of the spectrometer by a few degrees causes it to capture the different crop areas significantly separated from each other. In cases when the crop is not sufficiently homogeneous in the horizontal plane, the various points of a certain BRF correspond to the various crop areas, differing from each other in their geometrical parameters, and thus hindering data interpretation. A number of other technical difficulties exist which also lead to inefficiency in the passive remote sensing techniques used to measure the plant crop BRF in its hot spot. Therefore, an attempt has been made to develop a new technique of remote sensing of vegetation in this scattering region based on the utilization of the laser.

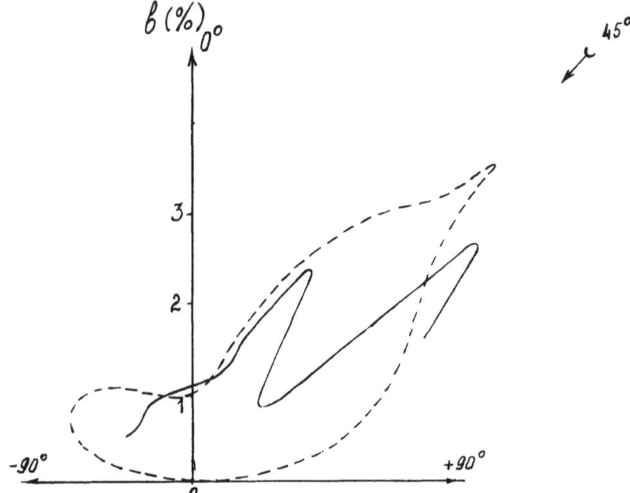

Fig. 28. The BRF of spruce fir plants: --- computer calculated and —— measured from the airborne platform: —— solar beam direction

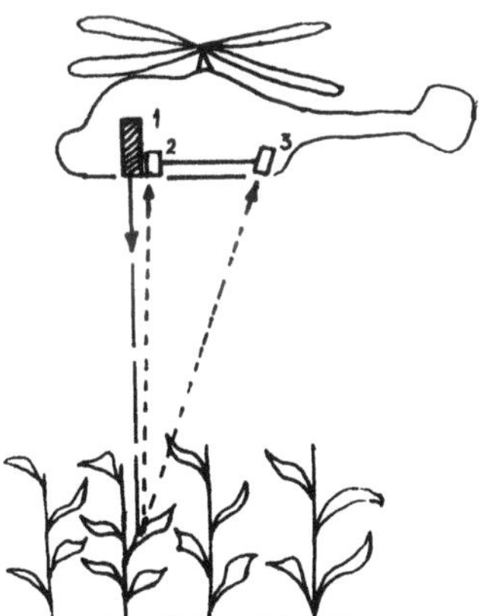

Fig. 29. Chart of laser remote sensing of vegetation from a helicopter using two spatially separated photodetectors. *1* Laser; *2* photodetector; *3* photodetector. --- Laser beam; ----- reflected beam

The laser remote sensing technique allows the reflective properties of the plant canopy to be measured in the hot spot region and allows the evaluation of the abrupt change in the vegetation BRF in this scattering region (Kanevski et al. 1985).

This technique is based upon measuring the laser-induced radiance reflected from plants by means of two spatially separated photodetectors. As can be seen in Fig. 29 one of these is placed close to the laser in order to observe the laser spot in the nadir which coincides with the laser beam direction. The other photodetector, placed at a certain distance from the first one, observes the spot at some angle to the nadir direction. The angle is ca. 3°–4° if the sensing is performed from a distance of 30–40 m and the distance between the photodetectors is 1.5 m (Kanevski et al. 1985).

The major aim of laser investigations is the determination of the relationship between the geometrical parameters of the vegetation canopy and its reflective characteristics in the hot spot region. The relation of intensities of the scattered laser radiation measured by two photodetectors placed at different sighting directions of the laser spot may be proposed as one of the informative parameters. The laser operates in a pulsed mode and the intensity relation measured for each radiation pulse does not depend on the variation in the leaf reflectance properties, but it does depend to a small degree on leaf orientation, for the difference in the spot sighting angles is very small and measurement of the radiation scattering intensity in the two chosen directions is being performed at the same point in the plant canopy per laser pulse. The relation being measured depends to a large degree on whether the laser spot is shaded by the leaves in the sighting directions of the second photodetector, which observes the spot at all acute angles to the nadir. Therefore, the mean value of this ratio, measured all over the experimental crop area, determines the peak value of the BRF for the two chosen scattering directions and is caused mainly by their specific spatial arrangement. The measured ratio is related to such geometrical characteristics of the plant canopy as leaf area index, number of leaves on a stem, and crop height.

The informativeness of the proposed technique might be substantially improved if the intensities of the reflected radiation pulse from each plant (for the two chosen scattering directions) could be detected and stored. If the results of such measurements were presented as intensity distribution functions of the number of reflected radiation pulses, certain intersecting relationships between the geometrical parameters of vegetation and the characteristics of the distribution functions could emerge. In this respect it would be useful to turn our attention to the results of remote sensing experiments (Kondratyev et al. 1987). A laser system with two spatially separated photodetectors was mounted on board a helicopter. The flights were operated at an altitude of 30 m, the distance between the photodetectors being 1.5 m. A 337-nm nitrogen laser with 5 kW output power and a pulse duration of 10^{-9} s operated at a repetition rate of 100 pps. The measurement and storage of the laser-induced scattering radiation intensities in the two chosen scattering directions and for each laser radiation pulse was accomplished by an optoelectronic system.

Crops in various stages of growth were chosen as the objects of investigation. Figure 30 depicts the intensity distribution functions of a wheat crop at the stage of milky ripeness. The presence of a specific peak in distribution function shows that in crops there are leaves oriented in the same manner, which is in agreement with the data concerning the geometry, as described in the literature (Ross 1981), of the wheat crop at this growth stage. The orientation of the scattering surface is one of the most significant parameters causing variation in the reflected radiation intensity. The decrease in the specific peak value in the distribution function, measured in the direction of the photodetector, indicates partial shading by the leaves in this direction. As shown in Fig. 30 the lodged crop has essentially changed the shape of the distribution function, making it resemble the equiprobable distribution of the pulse number according to their intensities, indicative of the absence of predominant leaf orientation in the lodged crop. The presence of predominant leaf orientation in the alfalfa crop serves to demonstrate the shape of the distribution function (Fig. 31). The increase in crop height from 0.15 to 0.50 m causes an increase in the distance between the distribution function peaks (corresponding to two sighting directions). The shift in the distribution function peak (corresponding to oblique sighting) in the direction of smaller values of the reflected radiation intensities, as the crop height

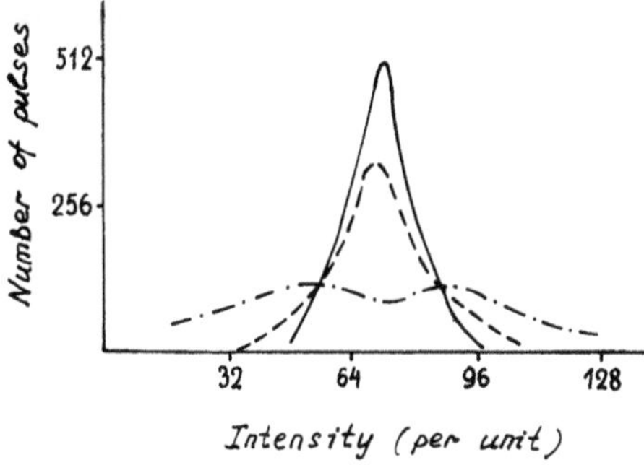

Fig. 30. Intensity distribution function of the reflected laser light pulses from the wheat crop. —— Backscattering; - - - - - angle scattering; –·–·– backscattering (40% of the crop is lodged)

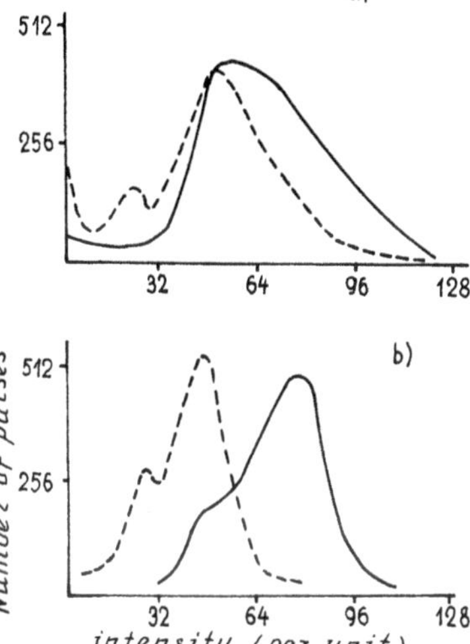

Fig. 31. The distribution function of intensities of laser-induced emission from an alfalfa crop. Backscattering direction (——); scattering at an angle of 3° (– – –). **a** The crop height, 10–15 cm; **b** the crop height, 60 cm

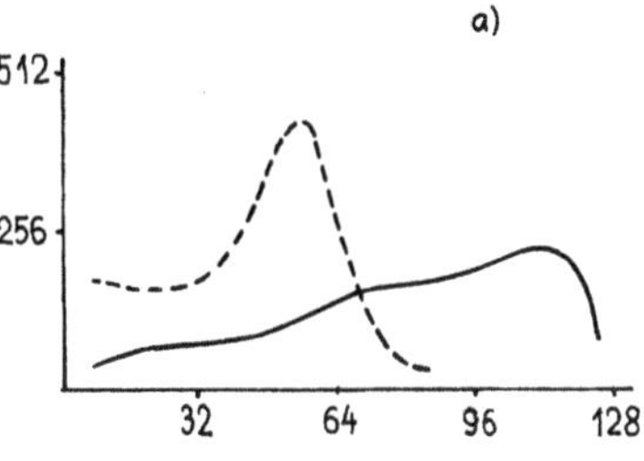

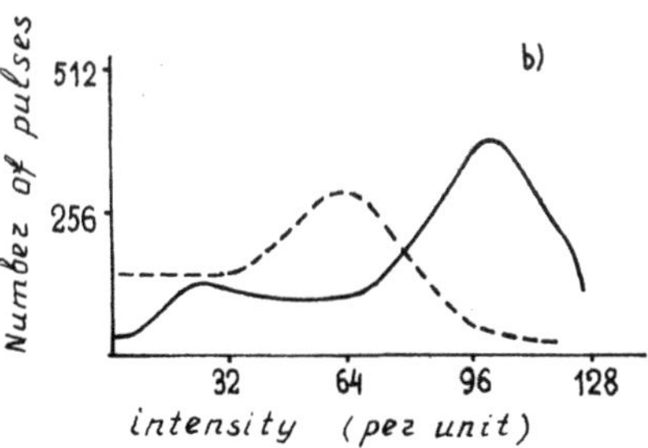

Fig. 32. The distribution function of intensities of reflected laser radiation emission from a corn crop. Backscattering direction (——); scattering at an angle of 3° (– – –). **a** The crop height, 70 cm; the crop density, 9 plants/m^2; **b** the crop height, 70 cm; the crop density, 16 plants/m^2

increases, points to the increased probability of the oblique sighting direction (of the second photodetector) being shaded by the leaves. The distribution functions obtained for the corn crop (*Zea mays*), shown in Fig. 32, are of special interest. The increase in the crop density from 9 to 16 plants/m^2 results not only in a variation in the distance between distribution function peaks, obtained at the two scattering directions, but also in the redistribution of the relation between the values of these peaks. The presented results demonstrate the usefulness of compiling a data bank for the hot spot area reflecting the characteristics of various plant canopies. This would make it possible to develop a technique for remote assessment of the geometrical parameters of vegetation which are closely related to growth stage, biomass, and other characteristics that are of importance for agriculture and forestry.

The above-described technique of laser remote sensing of the geometrical structure of vegetation may be developed further to increase the information obtained.

The further development of the above technique calls for a panoramic (surface) scanning of a small plant canopy area, where laser beam scattering occurs. The observation of the vegetation area to be investigated in this case should be performed at an acute angle to the laser beam direction, as shown in Fig. 33. The laser spot diameter should be comparable to the typical leaf dimensions. In such a case the image obtained through a panoramic photodetector would look like a set of spots with various diameters located at different distances from each other. The

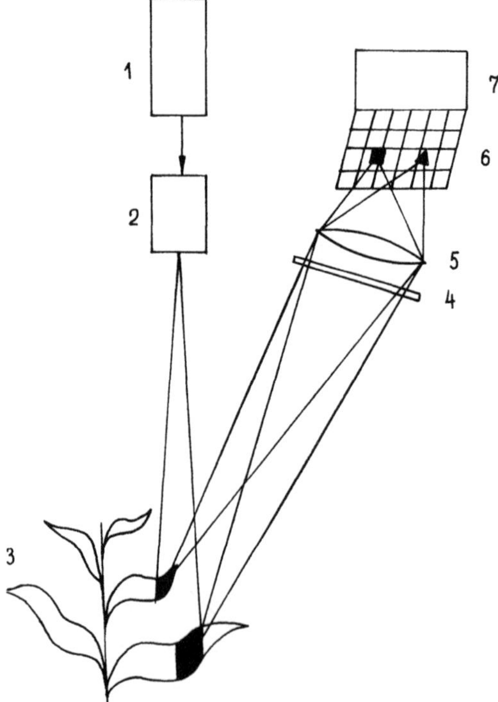

Fig. 33. Laser remote sensing of the vertical structure of the plant canopy. *1* Laser; *2* collimator; *3* plants; *4* filter; *5* lens; *6* photosensor, array; *7* image processing system

major mechanism necessary to form such an image is the splitting of the laser beam on the leaf surface blocking the laser beam inside the plant canopy, as shown in Fig. 33. The presence of an acute angle between the laser beam and the sighting direction of the panoramic photodetector does not allow a clear perspective observation of the splitting beam structure (Fig. 33). In fact, the development of a new technique to assess the vertical structure parameters of the vegetation was based on the above described situation. Useful data may be obtained through statistically treated images of the spatial structure of the laser beam in the process of splitting on the leaves of different layers of the plant canopy (Kanevski et al. 1989). Figure 33 shows the instrumentation chart used in the experiment. The vegetation was illuminated by a laser with a pulse repetition rate of 50 pps. The television image scanning was synchronized in phase with the laser pulse regime. Each frame was processed by a computing system which counted the number of spots in the frame and sorted them out according to their number. Afterwards, the probabilities of the laser beam splitting into two, three, four, or more parts were calculated. The results of the splitting are determined by the vertical structure of the plant canopy, i.e., by the number of its layers (Kanevski et al. 1988). The described technique of laser remote sensing of vegetation, giving the possibility to obtain data on the vertical structure of the vegetation, is only the first step toward the development of a new approach similar to the widespread use of computer tomography in medicine. Here, we deal with the foundations of laser remote tomography of the plant canopy.

4 Conclusions

The discussion of the results of applying lasers for remote measurements leads to the following conclusions. The use of lasers provides information of two sorts: (1) information similar to that produced by other methods; (2) unique information only made available by experiments with lasers. This includes data concerning the geometrical structure of crops and the spatial arrangement of their separate phytoelements. Of great interest is the possibility of studying parameters of the vertical structure of the plant canopy such as the number of vegetation layers and the vertical distribution of biomass. These experimental data are very important for determining the phase of crop development of specific cultures and the light regime in crops is important because their productivity is dependent on the amount of light absorbed. Until recently studies in this field were based on the use of mathematical calculations of hypothetical models. The use of lasers provides a reliable experimental foundation and this will, no doubt, lead to rapid progress in this field of science.

Remote measurements of the J_{685}/J_{735} ratio in the leaf fluorescence spectrum make it possible to determine their chlorophyll content. These data may be used to develop nonspecific tests of stresses or to measure the time profiles of plant pigmentation levels that enable one to predict the yields of grain crops (Kochubey et al. 1988). This type of information can also be obtained, as mentioned in Section 1, by using passive probing methods based on measurements of the parameters of light reflected from plants with high-resolution spectrometers (Kochubey et al. 1990). Comparison of two methods of remote measurement of chlorophyll content (the fluorescence method and the method using the form of the red edge of the reflection spectrum) shows that the accuracy of measurements is almost the same for both methods (about 50 mg m^{-2}) when the pigment concentration varied from 100 to 400 mg m^{-2}. Both methods are practically inoperative with concentrations above 500 mg m^{-2}. The fluorescence method needs more sophisticated and power-intensive instruments. This method, however, is advantageous compared to passive methods in remote measurements of objects with a low percentage of projected coverage, because the reflection from the soil greatly interferes with the measurements. Similar measurements using chlorophyll fluorescence are insensitive to this kind of interference.

We believe that the use of the blue-green luminescence of plant leaves for specific testing of stresses, particularly in combination with measurements of chlorophyll fluorescence, has good prospects. The development of these methods will also provide information that cannot be obtained by passive methods. However, this field needs further development. The combination of passive and active methods appears to be the most promising direction in the field of remote sensing of plants.

References

Barz W (1977) Degradation of polyphenols in plants and plant cell suspensium cultures. Physiol Veg 15: 261–277

Bauer ME (1985) Spectral inputs to crop identification and condition assessment. Proc IEEE 73, 6: 1071–1085

Belyaev BI, Kiselevsky LI, Smetanin EA, Pluta VE (1978) Portative fasting spectrometer MSS-2. Appl Opt 29: 1011–1018 (in Russian)

Billard JP, Boucard J (1982) Effect of sodium chloride on the nitrate reductase of *Suaeda maritima* var. *macrocarpa*. Photochemistry 21(6): 1225–1229

Blazej A, Suty L (1973) Raslinné fenolové z lúteniny. Vydavatol Tech Ekon Lit, Bratislava, 240 pp

Brayon A (1966) Seasonal changes of plant tissues self fluorescence. Science articles of Tartu University 185. In: Investigation on plant physiology and biochemistry, vol 2. 2nd Republic Conf Plant physiology and genetics, Tartu, pp 75–83

Bristow MPF, Houston WH, Measures RM (1973) Development of a laser fluorescensor for airborne surveying of the aquatic environment. NASA Conf on the lasers for hydrographic studies, Wallops Island, Sept 1973, SP-375, 1973, pp 173–195

Bunkin AF, Vlasov DV, Galumyan AS, Malcev DM, Slobodyanin VP (1984) Universal apparatus complex for the distance laser airborne sensing. Tech Phys 54(11): 2190–2200 (in Russian)

Bunkin AF, Vlasov DV, Mirkamilov DM (1987) Physical basis of laser remote sensing of the earth's surface. FAN, Tashkent (in Russian)

Canto de Loura I, Dubacq JP, Thomas JC (1987) The effects of nitrogen deficiency on pigments and lipids of cyanobacteria. Plant Physiol 83: 838–844

Chappelle EW, Wood FM Jr, McMurtrey JE III, Newcomb WW (1984) Laser-induced fluorescence of green plants. 1. A technique for the remote detection of plant stress and species differentiation. Appl Opt 23: 134–138

Chappelle EW, Wood FM Jr, McMurtrey JE III, Newcomb WW (1985) Laser-induced fluorescence of green plants. 3. LTF spectral signatures of five major plant types. Appl Opt 24: 74–80

Chetverikov AG, Richards GP, Peisahzon BI (1976) Application of the fluorescence spectroscope for vital observation of the redox state in drought plant cells. Dokl Acad Sci USSR 230: 492–495 (in Russian)

Chirkova TV, Dragunova EV, Bugrova MP (1977) Redox reactions of flavoproteins and pyridine nucleotides from roots of plants different in their resistance to oxygen deficiency studied in vivo. Plant Physiol 24: 126–131 (in Russian)

Collins W (1978) Remote sensing of crop type and maturity. Photogr Eng Remote Sens 44(1): 43–45
Duysens LNM, Amez J (1957) Fluorescence spectrophotometry of reduced phosphopyridine nucleotide in intact cells in the near ultraviolet and visible regions. Biochim Biophys Acta 24: 19–26
El-Sharkawy AM, Salama FM, Mazek AA (1986) Chlorophyll response to salinity, sodicity and heat stresses in cotton, rama and millet. Photosynthetica 20: 204–211
Estabrook RW (1962) Fluorimetric measurement of reduced pyridine nucleotide in cellular and subcellular particles. Anal Biochem 4: 231–236
Ferns DS, Zara SJ, Barber J (1984) Application of high resolution spectroradiometry to vegetation. Photogr Eng Remote Sens 50: 1725–1735
Frank E, Hoge RN, Swift K (1983) Feasibility of airborne detection of laser-induced fluorescence emission from green terrestrial plants. Appl Opt 22: 2991–3000
Gehlhaar V, Gunther KP, Lutker I (1981) Compact and highly sensitive fluorescence lidar for oceanographic measurements. Appl Opt 20: 3319–3320
Gerstl SA, Simmer C, Powers B (1986) The canopy hot-spot as crop identifier. In: Damen MCJ, Sicco Smit G, Verstappen HTM (eds) Remote sensing for resource development and environment management. Proc 7th Int Symp Enschede, 25–29 Aug 1986. Balkeme, Rotterdam, pp 261–263
Horler DNH, Dockray M, Barber J, Barringer AR (1983) Red edge measurements for remotely sensing plant chlorophyll content. Adv Space Res 2: 273–277
Il'in VP, Kochubey SM (1987) Changes in the luminescence spectra of wheat leaves by sulphate in the soil. Physiol Biochim Kult Rst 19: 258–262 (in Russian)
Il'in VP, Kochubey SM, Shelyah-Sosonko YuR (1986) Changes in plant luminescence in the regions of metal deposits. Dokl Acad Sci UK SSR Ser B 10: 51–53 (in Russian)
Kanevski VA, Ross JK (1983) A Monte Carlo simulation model for radiation conditions in coniferous trees. Acad Sci Estonian SSR, Div Phys, Math Tech Sci Preprint, Tartu, 32 pp (in Russian)
Kanevski VA, Ryasantsev VF, Perekrest ON et al. (1985) About the possibility of the remote sensed laser diagnostic of the crop condition using their luminescence characteristics. Sov J Remote Sens 6: 37–39 (in Russian)
Kanevski VA, Movchan YaI, Shelyag-Sosonko YuR et al. (1988) A remote sensed biochemical method for searching for mine deposits. Certificate of invention 1365009, Bull 1 (in Russian)
Kanevski VA, Ross JK, Fedak VS, Shelyag-Sosonko YuR (1989) A remote sensed way to study the plant objects and the equipment for this. Certificate of invention 1460625, Bull 7 (in Russian)
Karapetyan NV, Bukhov NG (1986) Chlorophyll variables of fluorescence as indicator of the plant physiological state. Physiol Rast 33: 1013–1026 (in Russian)
Karnaukhov VN (1978) Cell luminescence spectral analysis. Nauka, Moscow, 206 pp (in Russian)
Karnaukhov VN, Karnaukhova NA, Uashin VA (1984) The method and technique of luminescence cytodiagnostic. Preprint of Inst Biol Phys, Puschino (USSR) (in Russian)
Kim HH (1973) New algae mapping technique by the use of an airborne laser fluorosensor. Appl Opt 12: 1454–1459
Kochubey SM (1986) The pigment organization of the photosynthetic membranes as the basis of the energy security of photosynthesis. Naukova Dumka, Kiev (in Russian)
Kochubey SM, Shadchina TM, Odinoky NS (1986) Manifestation of nitrogen nutrition deficiency in plants by spectral characteristics of leaf fluorescence. Physiol Biochem Kult Rast 18: 35–39 (in Russian)
Kochubey SM, Kobetz NI, Shadchina TM (1987) The shape reflectance spectra of the leaves as informative basis of the remote sensing of crop state. Physiol Biochem Kult Rast 19: 539–545 (in Russian)
Kochubey SM, Shadchina TM, Kobets NI, Dmitrieva VV (1988) Correlation between reflectance characteristics of winter leaves and nitrogen and chlorophyll content in term during the vegetation. Physiol Biochem Kult Rast 20: 530–534 (in Russian)
Kochubey SM, Kobets MI, Shadchina TM (1990) Spectral properties of leaves as a basis of remote sensing research. Naukova Dumka, Kiev, 132 pp (in Russian)
Kohen E, Kohen C, Thorell B, Akerman L (1968) Kinetics of the fluorescence response to microelectrophoretically introduced metabolites in the single living cells. Biochim Biophys Acta 158: 185–188

Kondratyev KYa, Fedchenko PP (1982) Spectral reflectivity and recognition of vegetation. Gidrometeoizdat, Leningrad, 216 pp (in Russian)
Kondratyev KYa, Kanevski VA, Ross JK, Pozdnyakov DV, Ryazantsev VF, Fedchenko PP (1987) Laser remote sensing of vegetation. Acad Sci USSR, Leningrad, 168 pp (in Russian)
Lehninger AL (1972) Biochemistry. Worth, New York
Lichtenthaler HK (1987) Chlorophyll fluorescence signature of leaves during the autumnal chlorophyll breakdown. J Plant Physiol 131: 101–110
Lichtenthaler HK (ed) (1988a) In vivo chlorophyll fluorescence as a tool for stress detection in plants. In: Applications of chlorophyll fluorescence. Kluwer, Dordrecht, pp 119–132
Lichtenthaler HK (ed) (1988b) Remote sensing of chlorophyll fluorescence in oceanography and in terrestrial vegetation: an introduction. In: Applications of chlorophyll fluorescence. Kluwer, Dordrecht, pp 287–297
Lichtenthaler HK, Rinderle U (1988) The role of chlorophyll fluorescence in the detection of stress conditions in plants. CRC Crit Rev Anal Chem 19: (Suppl) 1: 29–85
McFarlane JC, Watson RD, Theisen AF et al. (1980) Plant stress detection by remote measurement of fluorescence. Appl Opt 19: 3287–3289
Measures RM, Houston HR, Stephenson DG (1974) Laser induced fluorescence decay spectra. A new form of environmental signature. Opt Eng 13: 494–450
Öquist G (1986) Effects of winter stress on chlorophyll organization and function in Scots pine. J Plant Physiol 122: 169–181
Penny MF, Abbot RH, Phillipe DM (1986) Airborne laserhydrography in Australia. Appl Opt 25: 2046–2053
Rachkulik VI, Sitnikova MV (1981) Reflectivity and state of vegetation cover. Gidrometeoizdat, Leningrad, 287 pp (in Russian)
Rinderle U, Lichtenthaler HK (1988) The chlorophyll fluorescence ratio F_{690}/F_{735} as a possible stress indicator. In: Lichtenthaler HH (ed) Applications of chlorophyll fluorescence. Kluwer, Dordrecht, pp 189–196
Ross J (1981) The radiation regime and architecture of plant stands. Junk, The Hague, 391 pp
Ross JK, Marshak AL (1988) Calculation of canopy bidirectional reflectance using the Monte Carlo method. Remote Sens Environ 24: 213–227
Safaraliev PM, L'vov NP, Mardanov AA, Kretovitch VL (1984) About reasons of decrease nitratreductase activity in beans upon soil salinity. Plant Physiol 31: 658–665 (in Russian)
Schmid PPS, Feucht W (1986) Kohlenhydrate, Chlorophyll, Prolin und Polyphenole in Blättern von Kirschkombination (*Prunus avium* L. auf *P. cerasus* L.) mit unterschiedlichen Streßsymptomen. Angew Bot 60: 365–372
Shadchina TM, Beloivan CA (1992) Testing of the plant stress on spectral parameters of its luminescence. In: Regulation mechanisms of plant physiology and genetics. Naukova Dumka, Kiev, 160 pp (in Russian)
Shibayama M, Akiyama T (1986) A spectroradiometer for field use. VI. Radiometric estimation for chlorophyll index of rice canopy. Jpn J Crop Sci 35: 433–438
Siano SA, Muthazasan R (1989) NADH and flavin fluorescence responses of starved yeast cultures to substrate. Biotechnol Bioeng 34: 660–670
Sud'ina EG (1964) Changes in biosynthesis and state of the chlorophyll upon deficit of the same elements. Ukrain Bot J 21: 36–44 (in Russian)
Tusov UB, Korvatovsky BN, Paschenko VL, Rubin LB (1980) On a question about the nature of the chloroplast fluorescence at 735 nm upon room and low temperature. Dokl Acad Sci USSR 252: 1500–1504 (in Russian)

5 Global Monitoring of Forests with Radar

M.G. Holmes and F.I. Woodward

1 Introduction

Until the 1950s, radar was used primarily to locate and track man-made objects such as aircraft and ships. As the technology improved, and requirement for alternative methods of mapping the Earth's surface grew, imaging radars were developed which could be used for military reconnaissance purposes. A major advantage of these radars was that they were not influenced by cloud cover because the long wavelengths could penetrate clouds and water in the atmosphere. An additional benefit from using radar was that it produced its own illumination, so the collection of images was not restricted to daylight hours. The advent of satellites led to the ability to cover even larger areas than aircraft in a given time. Although early satellites suffered from relatively poor resolution by aircraft standards, system developments have led to the production of high resolution images and the computer ability to process the high data rate which this requires. The use of radar for detecting natural and man-made objects is also long-established, but the capabilities of radar have been less exploited than those of the visible and near infrared wavebands. The potential advantages of imaging radar have been extensively researched in the last two decades, but it is only in very recent years that commitments to increase satellite radar imagery for the purpose of studying vegetation have been made.

The growing interest in radar remote sensing is reflected in the research programmes of the last decade and in the current space proposals of a number of countries. It is evident in more recent literature that an increasing interest is being expressed in the feasibility of microwave remote sensing as an alternative to the more 'traditional' visible and infrared methods in situations where the latter prove unsuitable, such as in overcast conditions. However, this view is not universal, and it is more probable that synergistic use of the visible, infrared and microwave wavebands will provide the optimal approach to monitoring forested areas.

Notable projects due to be undertaken before the end of this century include the USA's Earth Observing System (EOS). This will consist of a suite of sensing and imaging instruments, the aim of which is to understand and monitor the processes and interactions leading to changes in the Earth's environment over a period of at least 10 years, thus allowing continuity of data records. Synthetic Aperture Radar (SAR, see below) will be one such instrument, the use of which should be most profitable in conjunction with infrared data (e.g. using the High-Resolution Imaging Spectrometer – HIRIS) for forestry studies. EOS SAR will provide multi-frequency, -polarization and- incidence angle observations of the Earth.

The European Space Agency (ESA) launched its Earth Resources Satellite (ERS-1; sometimes referred to as ESA Remote Sensing Satellite) in 1991, the objectives of which include ocean and ice imaging as well as land observations using amongst others, a synthetic aperture radar imaging system.

In both arable crop and forestry studies it would seem that a more thorough understanding of the relationships between target and associated microwave back-scatter is being pursued. Thus the realization of the potentials of remote sensing is only becoming truly recognized as new instruments and techniques are devised and implemented to meet current needs.

2 Radar Systems

Radar is an acronym for radio detection and ranging. Radar was originally developed to detect an object and to indicate its range (distance) and direction (position). Radar uses energy in the microwave range of the electromagnetic spectrum (Fig. 1), the commonest wavelengths for remote sensing purposes being in the approximate range of 1 to 100 cm. Each wavelength has a designated code, the commonest being P, L, S, C, X and K. The amount of radiation scattered back from the target varies with the wavelength and, as will be seen, this property is used to delineate different types of target. Another useful attribute of radar systems is that much information can be gained from the polarization characteristics of the target. Most radars can transmit with vertically (V) and/or horizontally (H) polarized radiation and can receive in either of these modes. This gives four possible permutations, such as vertical transmit, vertical receive (VV), vertical transmit,

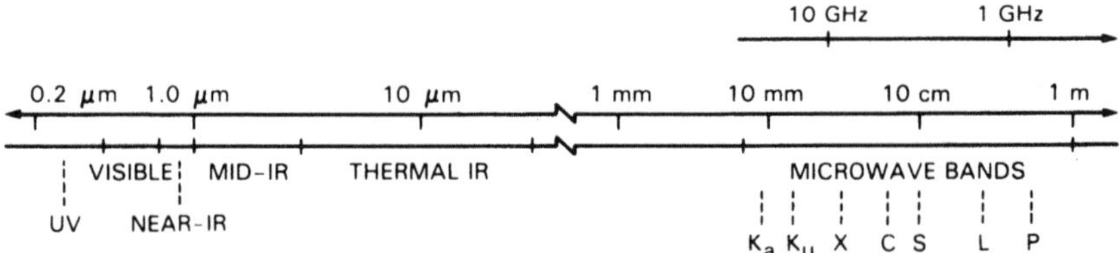

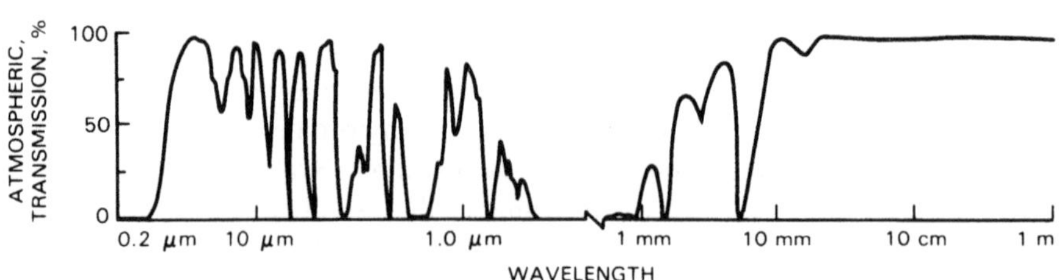

Fig. 1. Electromagnetic spectrum showing the atmospheric transmission windows (NASA 1987)

horizontal receive (VH) etc. Modern polarimetric radar systems measure the complete scattering matrix (amplitudes and differential phases). This allows synthesis of any polarization combination and optimization of reflected power or target/background ratio.

It is important at this stage to discriminate between active and passive sensors. Active microwave sensors provide their own source of energy and measure the backscattered return, whereas passive sensors measure the microwaves emitted by the observed target, such as soil. Active and passive microwave systems are sensitive to similar target characteristics, the main ones being the free water content (or dielectric constant) and geometry, be it in the form of surface roughness or spatial arrangement of vegetation. However, the present generation of passive systems do suffer from relatively poor resolution, even though geometric resolution can be increased in passive systems, as in active systems, by applying aperture synthesis (see Sect. 2.1). Passive spaceborne radiometer systems typically have a resolution of between 1 and 10 km, whereas a resolution of between 10 and 30 m is not an exceptional expectation from a modern active imaging system. While differing slightly in their information content, passive systems are suitable for global surveys, whereas active systems can be used for both global and detailed studies. This review restricts itself to active radar systems.

Active microwave systems function by transmitting pulses, or bursts, of microwave energy and then measuring the energy which is returned, or 'backscattered' from the target of interest. These active systems can be subdivided into two main forms. These are non-imaging systems (scatterometers), which transmit microwaves to the target and measure the returning energy, and imaging systems which operate in a similar manner but record an image of the scattered return from the target.

The majority of information on radar backscatter from vegetation has been derived from studies using airborne (as opposed to spaceborne) and ground-based radar systems. The data from these studies have led to the capacity to predict the optimum instrument and target parameters for identifying vegetation types and for monitoring changes in vegetation cover and soil moisture. It has been possible to derive a basic understanding of the physical parameters involved, and limitations of the systems, and the requirements for future developments in global monitoring of the vegetated land surface.

2.1 Imaging Radar

Imaging systems fall into two broad categories. These are real aperture radars (RAR), and synthetic aperture radars (SAR). A common feature of the RAR and SAR is that they are both a form of side-looking airborne radar (SLR). SLR functions by transmitting pulses of microwave energy to the side of the aircraft or satellite. The returning energy is then recorded digitally, with appropriate corrections being made for any change in speed of the aircraft. The SLR operates at an oblique angle and distortion of the image caused by the sideways transmission is corrected electronically to simulate an orthogonal, or vertical, view of the targets.

The early radar imaging systems (i.e. RARs) were limited by the physical size of the antennas because increased resolution could only be obtained by physically increasing the antenna length to impractical sizes. Design improvements led to the development of SARs. SARs provided a much improved resolution and greater flexibility with the ability to use longer wavelengths by electronically synthesizing a longer antenna length.

Calibration is an essential aspect of radar remote sensing. Radiometric calibration and geometric calibration are the two main forms which need to be considered here. Radiometric calibration refers to the accuracy and precision of the radar backscattering coefficient (σ°) measurement. Geometric calibration refers to the geometric accuracy of the position of a particular feature in a radar image relative to the position of that feature on a cartographic grid. The importance of good geometric calibration for registration of surface features for both SAR image analysis per se, and for comparison of SAR imagery with data from other imaging systems is obvious, but radiometric calibration needs further discussion.

Radiometric calibration incorporates both absolute and relative calibration. Absolute calibration determines the accuracy of the measurement of σ°; relative calibration determines the precision with which separate images, or different areas within a single image, can be compared. Calibration accuracies required for SAR studies of vegetation are commonly cited in the literature as being $\pm$ 1 dB for absolute, and $\pm$ 0.5 dB for relative radiometric calibration.

Accurate absolute calibration is important for several reasons. Some target parameters can be assessed directly if accurate measurements of σ° have been made. In forestry studies, there is evidence that not only leaf area index, but also the moisture content of surrounding bare soil can be accurately determined with an accurate radar of the appropriate frequency. The other main reasons for the desirability of accurate absolute calibration are that this not only allows comparison of data with data obtained by other instruments, but also permits accurate comparison of backscatter data obtained at different frequencies.

Good relative calibration with time (i.e. stability) is essential for monitoring vegetation. The ability to study temporal changes in a specific target can provide information on changes in backscatter resulting from disease, climate change, and cropping by man. Temporal changes can also be used to derive characteristic signatures for many species which have individual patterns of seasonal changes in backscatter. The spatial aspect of good relative calibration within an image is of particular importance when studies are being made which require high radiometric fidelity in an image, such as studies for tree classification.

Two main methods are used for radiometric calibration. The first involves internal monitoring of the instrument characteristics. By measuring the transmitted power, receiver gain, antenna gain, and the pointing direction, it is possible to relate the received power and the radar backscattering coefficient, σ°, by using the radar range equation. The second method is essential for forestry studies and involves using ground targets. These ground targets may be passive, such as corner reflectors with known backscatter cross-sections, or active, such as transponders.

2.2 Spaceborne Imaging Radar

The first spaceborne imaging radar was the L-band instrument launched on Seasat in 1978. Although it failed after a relatively short lifespan of 3 months, Seasat provided a large quantity of data. Most of the images were of oceans, but enough observations were made over land areas in the Northern Hemisphere to indicate the potential of spaceborne SAR for vegetation studies.

Seasat was followed by the Shuttle Imaging Radar-A (SIR-A) which was launched on Columbia in 1981. Like Seasat, SIR-A used 23 cm wavelength and only HH polarization. Unlike Seasat, SIR-A acquired images mostly over land, with emphasis on geological features.

SIR-B was carried into space by Challenger in 1984. This SAR used the same wavelength and polarization configuration, but was designed with the additional facility of an articulating antenna which meant that the data could be collected at a variety of incidence angles. The images obtained from SIR-B confirmed the ability of radar to delineate between different vegetation cover types, and the sensitivity of the system to variations in soil moisture.

The potential of spaceborne radar systems for mapping vegetation and soil (as well as the study of other geoscientific data such as ocean waves, ice cover, geological structures, land use patterns) led to the current development of new SAR technology for future space missions. Satelliteborne SARs which are active or currently projected are C-band ERS-1, L-band Japanese-ERS-1 (J-ERS-1), and C-band Radarsat. In addition, a multi-frequency, multi-polarization, variable incidence angle SAR is scheduled for two flights on Space Shuttles in the early 1990s. Known as SIR-C/X-SAR, this L-, C-, and X-band system has a variety of technical advances which are expected to provide major improvements in the information content of the images. In the more distant future, it is hoped that EOS will also support an L-, C-, and X-band SAR system.

2.3 *Advantages of Radar*

The mode of operation of the radar instrument, and the physical laws governing the propagation of microwaves in scattering media, introduce some unique characteristics for remote sensing of forests. The result is that radar imagery has two substantial advantages over conventional remote sensing using the visible and near infrared wavebands. The most important asset of radar is the use of its own energy to map the terrain. Whereas conventional methods (excluding thermal infrared and passive microwave sensors) rely on reflected sunlight for the production of an image, SLRs provide their own illumination with the result that surveys can proceed around the clock and are not restricted to daylight hours. In addition, SLRs are rarely hampered by the weather. The wavelength which are used can penetrate cloud and haze, and even light rain, although contrast can be reduced by wet surfaces. Problems only arise in extreme weather conditions such as rain storms. It is therefore much easier to plan radar studies and campaign timetables can usually be adhered to.

Radar possesses several attributes which can be exploited for global monitoring of vegetation and soil. For example, radar is very sensitive to the geometry of vegetation and the underlying surface, thereby allowing the possibility of species identification from its morphology and the extraction of information on the topography of the land on which it grows. Of equal importance is the ability to detect relatively small changes in the water content of the target. This not only provides information about the physiology and identity of the vegetation, but is also useful for assessing changes in the water content of the soil on which the productivity and development of the vegetation depends. An additional benefit derives from the variable extent to which different wavelengths are attenuated by plant canopies. As a result of this, it is possible to derive complex information on the layered structure of densely vegetated areas.

Active radar studies are by no means perfect. Even less is understood about the interaction of microwaves with soil and vegetation than is the case in the visible waveband. The objective of this work is to review the current state of knowledge of radar/target interactions and to emphasize those aspects which are relevant to monitoring of climatic effects on vegetation with a special emphasis on forests.

2.4 Combining Radar and Optical Studies

Although radar can be used around the clock and is only restricted by the most severe weather, observations with optical sensors are limited to clear-sky conditions. Combining radar and optical studies has the advantages of not only filling information gaps during overcast or hazy periods, but also of providing additional information which can be combined with optical data obtained under clear-sky conditions. Apart from improving our understanding of radiation/target interactions, and vegetation/atmosphere interactions, one of the most promising areas of combining radar and optical studies in a forestry context is in the improved tree classification rates which are possible.

No substantial information is available yet for forests, but a simulation of replacing missing Landsat data with radar data in a multi-date classification of crops is shown in Fig. 2. Here, the typical decrease in classification accuracy is demonstrated when data were not obtained on the first day of a multi-temporal study. However, if radar information is substituted for the missing Landsat data, the resultant correct classification is not only recovered, but is exceeded, but the use of radar data.

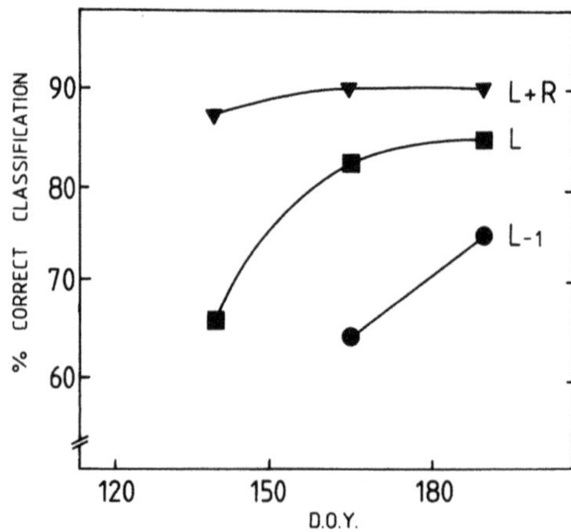

Fig. 2. Simulations replacing missing Landsat data with radar data in a multi-date classification. *L* Complete Landsat data; *L-1* 1 Day of Landsat missing; *L + R* substitution of missing Landsat with radar (After Ulaby, in NASA 1987)

The most exciting combination of radar and optical studies in the future is the integrated use of HIRIS (high resolution imaging spectrometer) in EOS. HIRIS will provide 196 spectral bands between 0.4 and 2.5 μm. Both HIRIS and SAR will provide images of similarly high resolution and therefore offer the potential for exploiting the different types of information provided by the two types of instrument. Although much more needs to be understood about the propagation of visible, IR and microwave radiation in vegetation canopies, the data provided by HIRIS and SAR should present an important opportunity to test compatible observations in current semi-theoretical and empirical models and to apply the conclusions to the monitoring of forests.

Apart from HIRIS and SAR, MODIS (moderate resolution imaging spectrometer) should also provide useful data which can be combined with SAR imagery. MODIS has a relatively low resolution which is only about 1/20th of SAR and HIRIS. Nevertheless, the ability of MODIS to monitor optical, shortwave- and thermal-IR radiation, and its relatively fast global coverage of 2 days, presents the potential for monitoring changes in land surface temperature over large areas. The most important information from MODIS will be the surface albedo and temperature data which can be combined with SAR data on land surface moisture content to provide an understanding of the land/atmosphere hydrologic processes.

3 Forestry Studies

3.1 Instrument Parameters

Several characteristics of the radar instrument influence the resulting image. The main instrument parameters or characteristics which are of importance are the frequency or wavelength, the polarization, and the incidence angle which is used. The extent to which radar wavelengths are attenuated or backscattered is influenced by all of these factors. For example, longer wavelengths tend to penetrate more deeply into vegetation or dry soil and therefore influence the characteristics of the returning radiation. This important feature, and the influences of polarization and incidence angle are discussed below.

3.1.1 Frequency

Field scatterometer and aircraft radar experiments have demonstrated in general that radar sensitivity to vegetation type is enhanced when the wavelength used is approximately equal in size to the canopy components (Fig. 3), which may be due, in part, to a resonance effect (NASA EOS Instrument Panel Report 1987). It follows, therefore, that by combining multi-frequency observations of a vegetation canopy, it might be possible to estimate morphological and biophysical parameters of that canopy. A combination of C- and X-band data may be more sensitive to crops, while L- and C-band data may be more useful for forest mapping. The inherent disadvantage of using only a single frequency is obvious and the potential for having multi-frequency satellite capability in the future cannot be ignored. EOS is one such programme where this will be implemented.

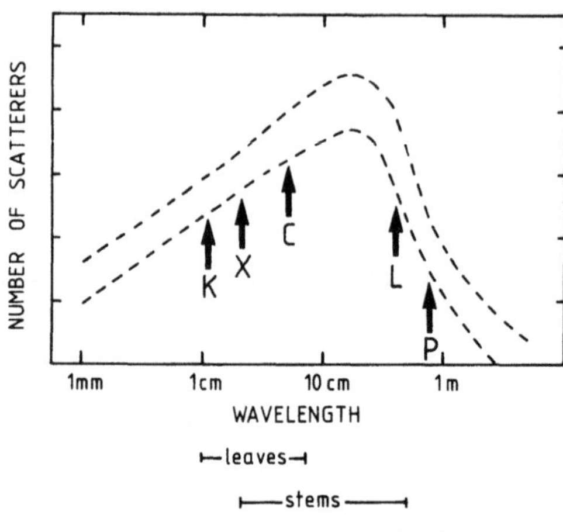

Fig. 3. Radar sensitivity as a function of the size of the canopy components and radar wavelength (NASA 1987)

The optimum frequency for analysis of forestry depends greatly on the application of the measurements. Early measurements with Ka-, X-, and L-band radar indicated that longer wavelengths were least suitable for delineation studies due to their inherent, reduced dynamic range (MacDonald et al. 1980) which the authors ascribed to the greater penetration of these wavelengths. The conclusion that shorter wavelengths are superior for forestry studies are supported by later studies (Rubec 1980; Bertholme 1983; Kessler et al. 1983; Horne and Rothnie 1985). There is strong indication that the optimum frequency is dependent on the specific delineation requirements. Both Goodenough (1980) and Lee (1980) conclude that both X- and L-bands could be used to map "clearcut" areas, although some preference was expressed for the X-band in HH polarization (Goodenough 1980). Recent analysis of SIR-B data (Wang et al. 1986) has demonstrated the sensitivity of L-HH imagery to surface features. It has been suggested that in conjunction with C-band, L-band would be most suitable for forest mapping.

If areas of forest regeneration or failure are to be analyzed, the L-band can provide more information than the X-band; Pala's (1980) work on SAR-580 data showed that whereas three subdivisions of regeneration could be classified with L-band, little useful information other than boundary identification could be obtained from equivalent X-band data. Pala concluded that coarse resolution L-band may satisfy general surveillance requirements. Difficulties were reported in the attempts to classify felled, recently planted, and non-woodland using data from the European SAR-580 campaign (Horne and Rothnie 1985). Similar difficulties have been reported in attempts to delineate between older and emergent crops (Le Toan et al. 1980; Knowlton and Hoffer 1981). Clearings and disturbances can be delineated with the aid of ground truth measurements using X-band SAR imagery (Churchill and Keech 1983).

The ability to delineate between woodland and non-woodland has been described with success for a range of wavebands. These include the K- (Daus and Lauer 1971), the X- (Parry 1974) and the L-bands (Le Toan et al. 1980). There is evidence from the European SAR-580 campaign that optically processed L-band data were superior to the X-band for tonal separation of woodland from other land cover classes (Horne and Rothnie 1985).

Areas of 5 ha or more can be discerned using Seasat satellite imagery (Hunting Geology and Geophysics 1981). Immediate conclusions are not straightforward, however, because failure to delineate woodland and non-woodland has also been described for both X- (Beaubien 1980) and L-bands (Beaubien 1980; Ulaby 1980). For boundary delineation, Horne and Rothnie (1985) found that X-band was superior (85% accuracy) to C-band (66%). It has to be borne in mind that radar technology is evolving rapidly and that some of the analyses for radar applications in the literature have not always taken into account limitations caused by saturation effects on the system and the relatively poor resolution of earlier instruments.

Species may be successfully interpreted, although there may be greater dependence on waveband for high accuracy than is the case for simply delineating woodland and non-woodland. Early successes were reported in the identification of species (Morain 1967; Peterson et al. 1969; both using the Westinghouse real aperture system). Later studies of SAR data by Knowlton and Hoffer (1981) conclude that X-band is best for delineation between coniferous and deciduous boundaries, though Wu (1986) was able to make the discrimination using C-band. Churchill and Keech (1983) and Churchill et al. (1985) were also able to discriminate between coniferous and deciduous woodland using X-band SAR-580 imagery; it is noteworthy, however, that Kessler et al. (1983) were unable to distinguish Douglas fir from oak even though they were using X-HH imagery from the same SAR-580 source.

Both Churchill et al. and Kessler et al. used texture for classification. It is possible that the different conclusions between the study groups resulted from the fact that there was a large age difference in the stands studied by Churchill et al., but negligible age difference in the forest types analyzed by Kessler et al. Ulaby et al. (1986), using L-band wavelengths, were able to improve the accurate delineation between broadleaf evergreen and broadleaf deciduous tree types using tonal and contrast differences in SAR images. They concluded that the intrinsic texture of such images is a valuable feature in discriminating between different land-use types. The subject of age discrimination is also discussed below in relation to texture analysis. Le Toan et al. (1980) were able to classify Maritime pine into three distinct age classes using L-band (both HH and HV polarization). Churchill et al. (1985) found that both the X- and C-bands (in both HH and HV polarization) were capable of providing images from which age discrimination in Scots and Corsican pine stands could be made.

Two relevant species delineation studies with satellite SAR are available. In the one, Wu (1983) claimed that it is possible to delineate three pine forest classes using L-band SIR-A data. In the other, although it was possible to discern woodland in general, it was not possible to discriminate between coniferous, deciduous and mixed woodland using Seasat imagery (Hunting Geology and Geophysics 1981).

Examples of woodland delineation from agricultural crops are shown in Fig. 4.

Stand density classification of Maritime pine is possible with the L-band (Le Toan et al. 1980). It is even possible to count single tree crowns of middle-aged to mature trees in woodland using stereograms of X- and C-band imagery from the European SAR-580 campaign, although poor illumination of lower storey trees

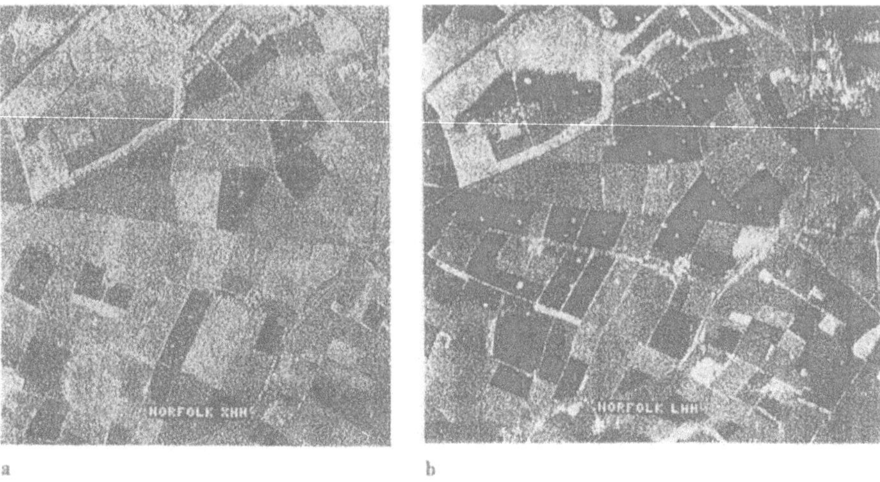

a b

Fig. 4. Delineation of woodland (very light tone in *upper left corner* of images) from various types of agricultural crops including barley, wheat, grass, carrots and peas. **a** X-HH; **b** L-HH (Courtesy of Hunting Technical Services)

necessitated the use of correction factors. The L-band has been reported as being superior to the X-band for tree identification in small tree clumps and for large individual trees (Horne and Rothnie 1985).

Although also dependent on incidence angle and leaf cover, the degree of canopy penetration is strongly dependent on the radar waveband which is used. Despite an early report that X-band penetrates further than L-band before deciduous trees are in leaf (Davis, 1973), there is substantially more evidence to support the conclusion that longer wavelengths penetrate deeper into a leafy tree canopy than shorter wavelengths. Shuchman and Lowry (1977), for example, determined that L-band penetrated further than X-band. This is due to the fact that the transparency of a vegetation layer increases with decreasing frequency. It is known that the amount of foliage on trees has a negligible effect on the backscatter of L-band radar, while X-band observations are governed mainly by the top layers of the canopy. This infers that factors other than foliage, i.e. branches, tree trunks, ground, influence the extent of L-band backscatter. As identical forest stands can produce different L-band signatures (MacDonald et al. 1980), the ground or standing water must be a major contributor to the backscatter, and is a factor which must be taken into account in forestry studies. As Richards (1987) points out, L-band would be a suitable tool to map flooding, since C- and X-bands are attenuated to a greater extent by the canopy. Clearly, meteorological conditions must be observed throughout an experimental time period, and a wavelength chosen that would either detect or ignore the presence of standing water within a forest, depending on the information required.

The greater penetration of long wavebands than short (Shuchman and Lowry 1977) is supported by Sieber (1985) and Hoekman (1987). The former concluded that the X-band was reflected from leaves, and that the L-band was reflected from branches. Using corner reflectors placed within a pine canopy, Churchill and Keech (1983) were able to determine that neither X- nor C-band radar was able to penetrate to (and return from) a depth of 6 m into a pine canopy. Richards et al. (1987) suggested the inclusion of four components into backscatter models of forest stands, having concluded that the trunk and ground together act as dihedral corner reflectors when using L-band radar. The C-band may provide a compromise between the higher and lower frequencies, but it is clear from the above discussion that more information is always available when multiple wavebands are used and Hoekman (1987) reiterates this point. He found that an increased percentage of backscatter originated from deeper within the canopy as the microwave frequency decreased (i.e. wavelength increased), concluding that by using a multi-band scatterometer (DUTSCAT), more information about the vertical distribution of backscatter, and hence the canopy itself, could be obtained.

3.1.2 Polarization

The penetration depth at different polarizations is influenced by the spatial orientation of the elements of the vegetation. A simplified description of this phenomenon can be demonstrated using corn (maize) as an example (Fig. 5). Here, horizontally polarized (HH) radar couples weakly to vertical stalks resulting in low attenuation. Vertically polarized (VV) microwaves, however, are attenuated to a greater extent causing a reduction in the penetration depth (Fig. 5). In this example, measurements using HH give information primarily about the lower canopy and the underlying soil medium, while VV data are related more to the upper canopy structure. However, this statement must be considered in the context of the wavelength used. As we have seen above, the penetration depth of microwaves depends on frequency such that the transparency of the vegetation layer increases with decreasing frequency. As a result, L-band observations are generally influenced by a greater depth of the vegetation canopy than X-band. Also, much vegetation does not exhibit such preferential orientation in its structure and, as a result, discrimination between polarizations may be impossible. It also has to be borne in mind that the backscattered signal does not depend on attenuation alone, but also on the scattering

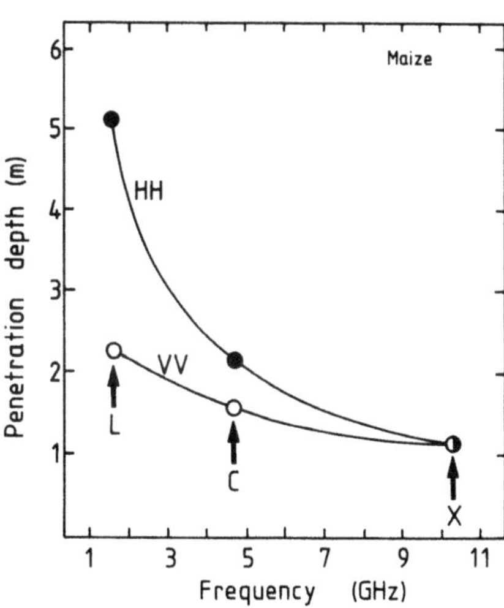

Fig. 5. Wavelength dependency for microwave penetration into a corn canopy. Penetration depth is defined as the depth at which the incident power is reduced to 37% (1/e) of the incident. The data presented are for an incidence angle of 40°. LAI = 2.8, plant height = 2.7 m, leaf volumetric moisture content = 0.65, stalk volumetric moisture content = 0.47 (After Ulaby et al. 1986)

properties of the vegetation which can induce biassed polarization effects on scattered radiation.

Fewer studies of polarization optima have been made than for waveband optima, but the conclusions which can be drawn are more clearcut in agricultural crops than in forestry studies. The degree of inhomogeneity of a surface or volume is strongly associated with the cross-polarization scattering coefficient of that surface or volume. The separation of vegetation types can be enhanced using cross-polarization data; for example, two vegetation canopies having similar geometries may have similar like-polarization backscatter, but it may be possible to separate them with observations from cross-polarization studies. Similarly, the distinction between bare soil and vegetation-covered surfaces is made easier using cross-polarization, due to the fact that the vegetation canopies depolarize the incident radiation more strongly than bare surfaces. The cross-polarization ratio (ratio of σ°_{HV} to σ°_{HH}) is the useful discriminating parameter in these studies. Earlier studies were based on higher frequency ($>$ 8 GHz) observations and led to the conclusion that little was to be gained by using cross-polarization (Bush and Ulaby 1977, 1978; Le Toan 1982). Lower frequency studies in the C-band by Paris (1982) demonstrated the advantage of using cross-polarization for delineation studies of corn and soybean which could not be achieved with like-polarization. The advantages of using cross-polarization are not restricted to lower frequencies. For example in crop studies, whereas Megier et al. (1985) concluded that cross-polarization was not as good as like-polarization in the X-band, they found that multi-polarization analysis (i.e. using both cross- and like-polarized data) greatly improved the ability to discriminate between crops.

Vegetation studies require polarimetric instrumentation in order to provide the maximum information. Because target structure influences the extent of radiation depolarization, there is a need to understand the dependence of depolarization on individual species structure. This infers that polarimetric studies need to be made of different species, and at different growth stages for each species, if the causal factors influencing depolarization are to be understood.

Generally speaking, vertically polarized radiation is more sensitive to the stalks of plants and tree trunks than that which is horizontally polarized, because of their vertical orientation. However, this will depend very much on the geometry of the forest stand in question. Although an early study by Peterson et al. (1969) concluded that cross-polarized radar was optimal for the identification of forest/non-forest and burned area boundaries, there is a larger volume of data which supports the conclusion that like-polarized measurements provide more information. This conclusion was reached by Goodenough (1980) and Knowlton and Hoffer (1981) who favoured HH polarized X-band for delineation between deciduous and coniferous forests, and by Wu (1983) who favoured VV polarization for forest classification. More recently, Wu (1987) noted that backscatter was higher with C-VV than C-HH for mixed hardwoods, but that C-HH backscatter was greater than C-VV for pine plantations.

A significant number of researchers have recently advocated the use of multi-polarization data in the study of microwave signatures from trees and forests. In particular, Evans et al. (1986) were able to relate multi-polarization images qualitatively to tree size, distribution and density, while Wu (1987) suggests that pine plantation biomass could be estimated using multi-polarization data. Wu and Sader

(1987) also suggest the use of such data in delineating a variety of surface features such as water-inundated wetland vegetation, as well as in characterizing forest vegetation (see also Hirosawa and Matsuzaka 1987). A study of three Japanese conifers by Hirosawa et al. (1987) led to the conclusion that needle-like leaves depolarize backscatter to a greater extent than non-needle-like leaves, such that the measured quantity of like- to cross-polarized backscatter was at its lowest with Maysu (Japanese pine) compared with Sugi (Japanese cedar) and Sawara (Japanese cypress). These results could be of value in the study of a range of tree species.

3.1.3 Incidence Angles

The majority of incidence angle studies have been restricted to correlative field observations and only a very limited number of studies aimed directly at understanding incidence angle effects have been made (de Loor et al. 1982). Optimal incidence angle for applied studies depends on the application. As with analysis of optimal polarization, the lack of comprehensive information for all wavebands limits interpretation. The need for more information is not only because such data will aid future monitoring programmes, but also because the data are required to aid model systems.

The influence of incidence angle depends on the polarization of the radar and the canopy orientation under examination. If a fully grown corn canopy is considered, using L-band radar, an increase in incidence angle from 0° to 90° has little effect on the penetration depth of HH-polarized radiation due to its low attenuation, whereas it decreases with VV polarization as incidence angle increases (Fig. 6). This phenomenon can be used to choose the most suitable incidence angle and polarization, depending on the application and information required.

The main influence of incidence angle is on the apparent roughness of the target, although the influence of surface roughness is reduced at certain angles where other factors, such as soil moisture, become more important. More information is required

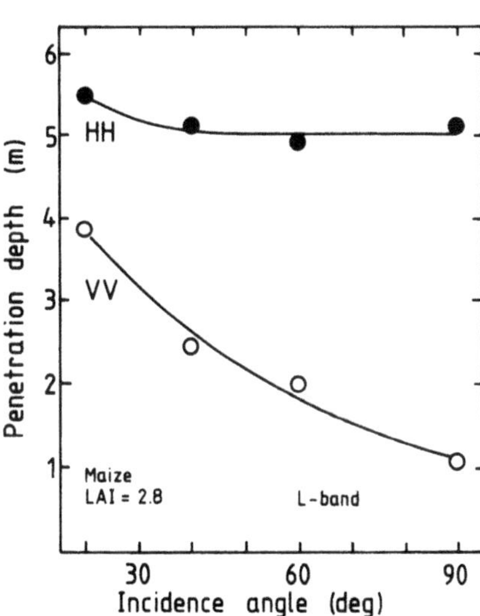

Fig. 6. Polarization and incidence angle dependency for microwave penetration into a corn canopy. Penetration depth is defined as the depth at which the incident power is reduced to 37% (1/e) of the incident. LAI = 2.8, plant height = 2.7 m, leaf volumetric moisture content = 0.65, stalk volumetric moisture content = 0.47 (After Ulaby et al. 1986)

on the evaluation of optimal incidence angle for woodland studies. Broad recommendations have been made to the effect that ground images should be carried out in the near, mid, and far ranges (Hardy et al. 1971; Lewis 1971). Rubec (1980) considered that shallow incidence angles were best for land classification. Wu (1983) concluded from X-band images that variations in incidence angle between 40° and 58° resulted in no detectable difference in the image. Vertical irradiation will provide greater penetration through a forest canopy for longer wavelengths because low angles of incidence result in scattering by tree trunks (EASAMS 1973). The effect of a change in incidence angle will therefore depend on the wavelength used. Hoekman (1987) demonstrated that at C-band and a low incidence angle (16°) about 40% of backscatter from poplar stands originated from ground and understorey, while at high incidence angle (60°) about 10–15% of the backscatter came from the ground and understorey. At L-band and low incidence angles, 75% of the backscatter came from the ground, implying that the poplar crowns are very transparent at these frequencies. However, as incidence angle increased, the relative backscatter contribution from the lower layers decreased, due to the radar 'seeing' more of the canopy at these oblique angles, as with C-band observations.

The advantage of using multiple incidence angle data is discussed by Cimino et al. (1986), who used three incidence angles, 33°, 44.6° and 59.4°, creating a colour composite image of an Argentinian forest system, as part of the SIR-B series of experiments. It was shown that forest types responded differently with respect to their brightness signatures at different incidence angles. For example, at low incidence angles, Lenga showed the strongest backscatter signal, which allowed the distinction between this and another forest type (Nire) to be made. It would be possible, therefore, to discriminate between forest species and various structures of a single species using multiple incidence angle radar imagery.

Ground topography plays a major role in determining radar backscatter (Hardy et al. 1971; Parry 1977). Such effects can lead to difficulties in delineating stands of deciduous and coniferous forest (Beaubien 1980), or even cause differences in the image of identical stands (Daus and Lauer 1971). This effect of topography is not surprising because topographical effects are effectively changes in incidence angle for backscattered wavelengths which penetrate to, and return from, the ground. In addition, changes in topography is likely to result in changes in the number of scatterers in the resolution cell.

From what is known of canopy penetration, it is to be expected that penetrating (i.e. longer wavelength) bands will be influenced strongly by ground topography, while bands which are strongly scattered by leaves (i.e. shorter bands) will be influenced by canopy surface topography. Multi-waveband knowledge of the fundamental relationships between incidence angle and backscatter would lead to an understanding of the influence of topography on backscatter signal. Multi-temporal, multi-waveband studies of a deciduous canopy would produce information from which wavelength-dependent topography effects could be analyzed.

3.2 Vegetation Parameters

A wide range of parameters affect the backscatter of microwaves from soil and vegetation. In a recent study undertaken for the European Space Agency (ESA), for example, it was concluded that a minimum of 4 instrument parameters, 28 vegetation parameters, 13 soil parameters, and 12 environmental parameters needed to be included in a data base which was designed for interpreting the mechanisms controlling backscatter from vegetation (ESA 1987).

Some generalizations can be made. The important characteristics of the radiation are the frequency, the polarization, and the incidence angle. The crucial features of the target in determining the proportion of radiation returning to the instrument are the biomass, the dielectric constant (which is closely linked to the water content), and the geometry of the vegetation. It therefore follows that the backscatter is dependent on the gross morphology of the vegetation which in turn depends on species and physiological age. One of the properties of radiation in the microwave waveband is that its ability to penetrate into the target (which is wavelength-dependent) provides a potential for studying not only the surface vegetation, but also the vegetation and soil below. In the case of soil, the predominant factors affecting backscatter are the morphology (roughness) and the dielectric constant (water content).

3.2.1 Penetration Depth

The depth to which microwaves penetrate into a vegetation canopy depends primarily on the wavelength of the radiation and, if the vegetation exhibits a preferential orientation, on the polarization of the radiation. The dependence on wavelength was seen in Fig. 5 for the penetration of L-, C-, and X-band microwaves at an incidence angle of 40°. This example demonstrated clearly the general phenomenon that penetration decreases with increasing frequency. An important consequence of the wavelength-dependence of penetration depth is that longer wavelengths can provide information on the entire canopy, whereas shorter wavelengths are useful for studying the upper layer of canopies.

Polarization differences in L-band penetration depth in a corn canopy were seen in Fig. 6. Vertically polarized radiation was much more strongly attenuated than horizontally polarized and there was a marked increase in the attenuation of the vertically polarized radiation with increasing incidence angle. Horizontally polarized radiation, on the other hand, penetrated deeply into the canopy and its attenuation showed negligible dependence on incidence angle. An important practical aspect of this type of observation is that vertically polarized data provide information which is predominantly related to the physical characteristics of the canopy, whereas the horizontally polarized radiation is largely penetrating the canopy and therefore providing information on the soil below. As pointed out earlier, however, the situation in vegetation which does not show a preferentially vertical orientation is not so straightforward and much more research is required.

3.2.2 Tree Morphology

The importance of identifying the dominant backscatter source from a forest stand is evident from recent literature, with a number of studies dedicated to this question. Tree geometry is a consideration in backscatter modelling that cannot be overlooked, since the microwave return is composed of scattering from various regions within the canopy. Richards et al. (1987), using L-band radar, have suggested that four components should be included in developing such models, these being volume scattering from the canopy, diffuse scattering from the forest floor, scattering from the canopy to the floor followed by specular reflection to the radar receiver, and scattering from the trunks to the floor followed by specular reflection to the radar receiver.

Zoughi et al. (1986) identified sources of backscatter and attenuation from various trees and shrubs at X-band wavelengths. For pine, this primary source was the needles, with cones more important for backscatter than attenuation. For other species (pine, oak, sycamore and sugar maple), leaves were the main cause of backscatter and attenuation. The radar return from a dense shrub, such as creeping juniper was shown to come from the top of the plant.

The formation of backscatter models is central to the effective use and understanding of microwave signatures, which has led Lamont et al. (1987) to emphasize the need for the development of techniques to measure and monitor plant structure and identify the dominant scattering components within a canopy. Such information could then be incorporated into these models. Paris (1986) and Bouman (1987) have shown, using short wavelengths, that radar return is sensitive to leaf size. Bouman, whose study was centred on agricultural crops rather than trees, concludes that longer wavelengths are less sensitive to changes in vegetation geometry than shorter ones, particularly with respect to beet and potatoes.

Microwave backscatter from vegetation is strongly dependent on the size of the scattering elements. This has been seen in Fig. 3, where the size of the elements which cause maximal backscatter varies with the wavelength of the radiation. In most instances, the greatest response is shown to scatterers which are of a similar size to the wavelength. This is one of several vegetation characteristics which contribute to the ability of multi-frequency imagery to discriminate between targets.

In addition, vegetation usually exhibits preferential orientation in its geometry. In most instances these orientations can be divided into preferentially vertical or preferentially horizontal. This results in polarization-dependent differences in the penetration and return of microwaves from the canopy, which in turn permits analysis of the canopy using different instrument polarization configurations. Useful approaches here are to use colour composites of multi-frequency images with each frequency shown in a different colour, or to look at the phase difference between horizontally and vertically polarized imagery.

Three detailed studies specifically aimed at understanding the relative contributions of forest structural elements to backscatter have been carried out. Westman and Paris (1987), using C-band, argue that leaf surface area is of more significance than biomass with respect to scattering properties. The conclusion that the effect of canopy moisture on backscatter is enhanced by dividing the total leaf mass into more scattering units (i.e. leaves) is consistent with Paris (1986), where parallel experiments on corn showed that radar return was sensitive to leaf size.

Attenuation through a forest canopy depends on the orientation of gaps with respect to the sensor and the properties of the various plant structures through which the radiation passes. Hoekman (1987) generalized the attenuation properties of two forest types at three wavebands (X-, C- and L-band). He stated that at X-band, deciduous tree crowns show high attenuation, while that of coniferous tree crowns is relatively low. It would appear, therefore, that at this wavelength, canopy architecture (i.e. the presence of gaps) determines the contribution of scatterers from the understorey and forest floor. At longer wavelengths, the backscatter originates from deeper within the forest structure.

The detection of forest cover on the basis of differences in signal caused by height has been reported for unusually high trees such as Douglas fir (Hardy et al. 1971) and mahogany (Sicco-Smit 1974). By using nadir-looking airborne radar, Bernard (1987) has shown the backscatter signal to be correlated with mean tree characteristics, such as height or density, which could be useful in forestry for monitoring tree maturity. Since there is a relationship between SAR image tone (which is proportional to backscatter) and tree age and height, and because total tree biomass is related to age, it follows that the digital number may be related to biomass. It has to be borne in mind, however, that the number of trees per unit area generally decreases with age in both natural populations and in managed forests. Wu's results (1987) showed that multi-polarization data are highly related to pine plantation biomass, and may therefore be useful in biomass estimation. Hoekman (1985) has noted the relative ease with which species in Europe can be differentiated using only one frequency (X-band), but whether this is a morphological effect is unclear. Species (i.e. morphology and physiology) and age are obviously involved, so morphological effects are clearly implied.

3.2.3 Vegetation Dielectric Properties

The dielectric constant of vegetation, as with soil, is one of the most important factors in determining radar backscatter. Along with the volume fraction and the geometry of the vegetation, the permittivity and loss factors are essential for an understanding of the attenuation of microwaves by vegetation. Unfortunately, measurements of dielectric are not straightforward and, largely for that reason, are not common in the literature although some measurements have been made (e.g. Ulaby and Jedlicka 1984). As with soils, the dielectric constant of vegetation increases with increasing water content. However, the limited information available indicates that the relationship between vegetation dielectric constant and frequency is more complex.

3.2.4 Multi-Temporal Studies

Radar backscatter from most vegetation types varies during the growing season as major factors influencing backscatter, such as water content and geometry, change. This progressive change in backscatter will vary between vegetation species according to the specific temporal and morphological characteristics of that species. Although the information available for forests is very limited (Hoekman 1985), there is a clear indication from the few studies which have been made with crops (Hoogeboom 1982, 1983) that multi-temporal imagery greatly increases the classification rate of several vegetation types including corn, milo, soybean, wheat and alfalfa (Bush and Ulaby 1977, 1978; Simonett 1978; Li et al. 1980; Ulaby et al. 1981; Shanmugan et al. 1983). Bush and Ulaby (1978) point out that crop classification using radar is time and site specific, and suggest that multi-date data acquisition is necessary if classification is to be more than 90% successful. The study of Brisco and Protz (1980) represents a rare exception to this general finding.

The revisit requirement for forestry studies is less stringent than for herbaceous vegetation. Closer-spaced measurements would probably provide more rewarding information. In the case of deciduous forests, there is a marked change in effective cross-section during spring (i.e. during leaf growth) and during autumn (i.e. during leaf abscission). Also, immediate recognition of changes in growth rate caused by preventable factors such as disease is of prime importance. Whereas such effects require a longer time-scale to manifest themselves in colder months due to lower metabolic rates of both tree and disease-causing organisms, the effects are relatively rapid during warmer periods which occur during leaf growth and in summer when deciduous trees accumulate dry matter.

Evergreen (mainly coniferous) tree species do not exhibit the same extent of seasonal change in radiation cross-section because there is a smaller change in leaf area. However, disease is a seasonal effect which progresses most rapidly during warmer seasons. The amount of information on the effect of disease is very limited, although analyses of the potential of SAR data for the detection of insect infestations in forests have been carried out (Goodenough 1980). Substantial multi-temporal studies are needed because a solid basis for understanding seasonal changes in backscatter from forests is required. Similarly, it is crucial that multi-wavelength and polarization experiments are performed, since backscatter is a function of radiation frequency and polarization, as well as season. The use of only one wavelength obviously restricts the amount of information which could otherwise be obtained.

One of the most interesting temporal effects at the moment is the influence of daily changes in the dielectric coefficient of vegetation on backscatter. McDonald et al. (1990) have recently reported a diurnal change in the dielectric coefficient of the trunks and branches of walnut trees which are paralleled by diurnal variations in backscatter at L-band. Similar observations have been made at C-band for both pine and beech (Holmes et al. unpubl.). This dynamic aspect of vegetation will require greater understanding if the 24-h capability of radar is to be proven reliable.

3.3 Soil Parameters

It has long been known that soil can play an important role in backscatter measurements of vegetation. The entire scope of the parameters influencing backscatter from soil is not understood, but soil dielectric constant, soil moisture content, and soil roughness and texture are known to be three of the most important factors.

3.3.1 Soil Dielectric Properties

The dielectric properties of moist soils are major factors in determining the microwave scattering and absorption by a soil (e.g. Dobson and Ulaby 1986). There have been many experimental studies of the dielectric behaviour of soil–water mixtures and several attempts have been made to model the dielectric behaviour (Hipp 1974; Hoekstra and Delaney 1974; Wobschall 1977; Wang and Schmugge 1980; de Loor 1983; Dobson et al. 1985; Hallikainen et al. 1985). Although none of

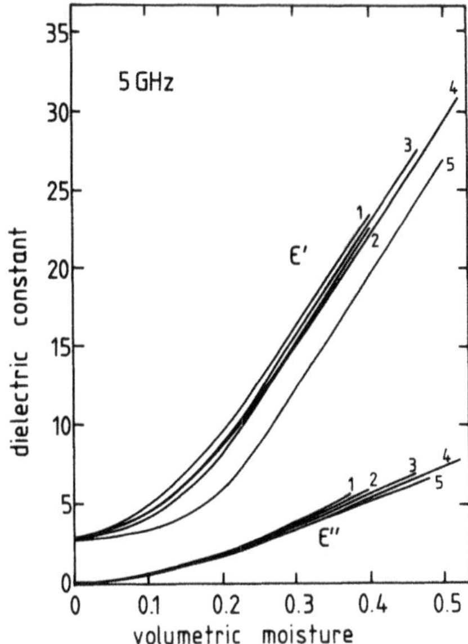

Fig. 7. Dependence of dielectric constant on soil type for a range of volumetric moisture contents. *1* Sandy loam; *2* loam; *3, 4* silt loam; *5* silty clay (After Ulaby et al. 1986)

the existing models for soil–water mixtures is capable of fitting experimental data on the basis of the soil physical parameters which have been used, it is nevertheless clear that the dielectric constant is a complex function of the soil–water content, the frequency of the radiation, and the soil type, as well as the temperature and the salinity of the soil–water mix. The water content, frequency, and soil type deserve special mention here.

Soil is a heterogeneous mixture of soil particles (the host material) and water. Dry soil exhibits a narrow range in real dielectric constant, or permittivity (ε'), between about 2 and 4, whereas that of water is about 81. The permittivity of soil is therefore a function of the water content of the soil and permittivity increases with increasing water content (Fig. 7). In nature, the ε' range of soil is from about 3 to about 30. Also, both ε' and ε'' (the dielectric loss factor) vary with frequency, the permittivity increasing with increasing wavelength, and the loss factor decreasing with increasing wavelength. This frequency dependency is largely a function of the water, rather than the soil moiety. The backscatter increases with increasing water content of the soil.

The amount of water associated with the soil is strongly dependent on soil type. The water held by the soil consists of two main types, bound and free. Bound water is held tightly to the soil particles by matric and osmotic forces. Free water, on the other hand, is a few molecular layers away from the particles and is at a sufficient distance from them to allow relatively free movement within the soil. The amount of bound water depends on the particle surface area, which in turn depends upon the particle size distribution of the soil. The dielectrics of equal amounts of bound water and soil, and free water and soil, are very different, with the result that the soil particle size and type produce a marked effect on the soil dielectric. Referring to Fig. 7, it can be seen that there is a small variation in ε' with different moisture-containing soil types, although there is no detectable difference in ε' for the dry soils.

It is important to note that the dielectric does not vary much with soil type per se, but that it varies with soil moisture content, which is a function of soil type.

The effects of soil organic matter content and the mineralogy of the soil are largely unexplored (Dobson and Ulaby 1986), but their direct effects on the retentive capacity of the soil for water cannot be ignored. A similar argument holds for the related parameter of field capacity. All three parameters (soil organic matter content, mineralogy, and field capacity) have been fairly extensively covered in ground measurements in both Europe and the United States.

3.3.2 Soil Moisture

Soil moisture can affect the ability to delineate between different types of vegetation (Shanmugan et al. 1983). This characteristic provides radar with a major advantage over other forms of remote sensing when information on climatic effects on soil moisture availability is required. Intensive scatterometer studies at the University of Kansas have indicated that the soil moisture content of bare soil can best be analyzed with the C-band and that HH polarization and incidence angles in the range 7°–17° gave optimum data for analysis if the confounding effects of roughness were to be minimized (Ulaby and Batlivala 1976). The same group found that similar instrument characteristics were optimal when the soil is covered by vegetation. They observed a strong correlation between backscatter and soil moisture content of the upper few centimetres of soil when it was planted with wheat, corn, milo, or soybeans (Ulaby and Bush 1976; Ulaby et al. 1979; Bradley and Ulaby 1981).

Brown et al. (1984) also observed a correspondence between wet areas (detected by aerial photography) containing no vegetation and C-band SAR backscatter at incidence angles between 30° and 45°. Similar conclusions have been reached by other workers (Schmugge 1980; Bernard 1981; Bernard et al. 1982; Le Toan et al. 1981; Le Toan 1982). Anomalies do exist; for example, the airborne scatterometer studies of Paris (1982) using C-band, HH polarization and 10° incidence showed significant effects of the vegetation rather than the soil. Clearly, as has been seen above, the relative extent to which vegetation and soil influence the backscattered signal is strongly dependent on the characteristics of both the instrument and the vegetation target. Wang et al. (1986) concluded that soil moisture could be mapped on an individual field basis, as well as within a field using SIR-B imagery at L-band wavelengths and an incidence angle of 21°. Nevertheless, it is clear that under some circumstances soil effects can be almost over-riding in determining the radar signal (Le Toan et al. 1981, 1984; Le Toan 1982).

Soil bulk density has been extensively recorded as a potentially important parameter for soil microwave studies. The influence of bulk density per se cannot be pin-pointed, it being linked in particular to the water content of the soil. Nevertheless, it is directly or indirectly an important soil property influencing the radar backscatter response (Ulaby et al. 1978; Schmugge 1980; Dobson and Ulaby 1981; Bernard et al. 1982; Wang 1983) because it is related to the relative proportions of soil (low dielectric) and water (high dielectric) which determine the composite dielectric coefficients.

One of the main problems with the analysis of soil moisture effects on the backscatter of microwaves is the influence from variations in roughness of the vegetation and the soil. One approach to overcome this problem has been to

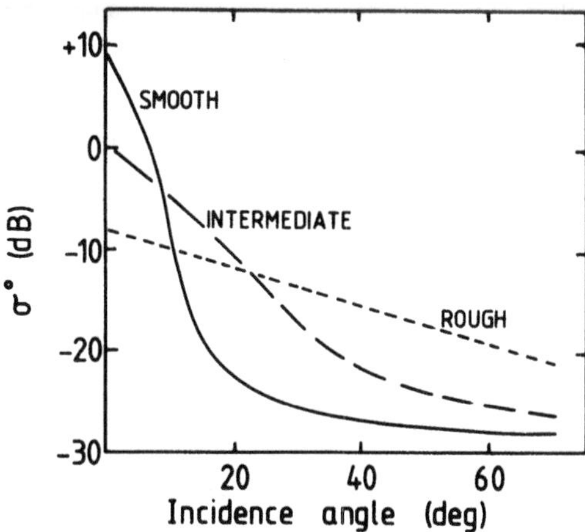

Fig. 8. Effect of soil roughness on the relationship between backscattering coefficient and incidence angle at 1.1 GHz (After Ulaby et al. 1986)

separate the data into groupings with nearly constant roughness and vegetation parameters. By this means it is possible to compare bare versus vegetated and smooth versus rough (Newton 1977; Ulaby et al. 1978, 1979; Newton and Rouse 1980; Njojku and O'Neill 1982; Wang et al. 1982; Taube and Theiss 1984).

3.3.3 Soil Roughness

The influence of soil roughness on the return signal needs to be understood and accounted for (Ulaby and Batlivala 1976; Schmugge 1978, 1980, 1983; Koolen et al. 1979; Ulaby et al. 1979; Chang et al. 1980; Schmugge et al. 1980). A greater understanding of the effects of soil roughness on the backscatter signal would greatly improve image analysis. Scatterometer measurements (Fig. 8) have demonstrated that the effects of roughness are minimized at incidence angles of about 10°. These data were obtained over bare soil using a range of roughnesses which are commonly found in agricultural fields. Another notable feature of these observations was that the effects of different roughnesses were least with shorter wavelengths. This is because all the surfaces are relatively rough for shorter wavelengths.

The row direction resulting from soil tillage has substantial effects on the angular dependence of radar backscattering. These periodic components of soil surface roughness are related to the analysis of the soil moisture effects because they are known to have substantial influence on the angular effect and look-direction effect of microwave backscatter (Ulaby and Bare 1979). Thus, even when soil roughness is approximately the same, tilled and untilled fields cannot be treated as being the same. The radar signal is most sensitive to azimuthal viewing changes (Ulaby and Bare 1979; Bradley and Ulaby 1981; Ulaby et al. 1982; Blanchard and Chang 1983). Elevation, frequency and polarization dependencies are also recognized (e.g. Dobson and Ulaby 1986) and more information on these influences is still required.

References

Beaubien J (1980) Forestry mapping using SAR. Intera Env Consult ASP-80-1

Bernard R (1981) A C-band radar calibration for determining surface moisture. Cent Rech Phys Environ (CRPE), Rue du General Leclerc, 192131 Issy-les-Moulineaux, France

Bernard R (1987) Nadir looking airborne radar and possible application to forestry. Remote Sensing Environ 21: 297–309

Bernard R, Martin P, Thony JL, Vauclin M, Vidal-Majar D (1982) A C-band radar calibration for determining surface soil moisture. Remote Sensing Environ 12: 189–200

Bertholme E (1983) Optical SAR-580 data for land use studies; an evaluation for S. Belgium. Eur SAR-580 Invest Prelim Rep (ESA)

Bouman B (1987) Radar backscatter from three agricultural crops: beet, potatoes and peas. Cent Agrobiol Res (CABO)

Bradley GA, Ulaby FT (1981) Aircraft radar response to soil moisture. Remote Sensing Environ 11: 419–438

Brisco B, Protz R (1980) Corn field identification accuracy using radar airborne imagery. Can J Remote Sensing 6: 15–21

Brown RJ, Guindon B, Teillet PH, Goodenough DG (1984) Crop type determination from multitemporal SAR imagery. Proc 9th Can Symp Remote Sensing. St. John's, Newfoundland

Bush TF, Ulaby FT (1977) Cropland inventories using satellite altitude imaging radar. Univ Kans Cent Res, Lawrence, Kans, RSL Tech Rep 330-4

Bush TF, Ulaby FT (1978) An evaluation of radar as a crop classifier. Remote Sensing Environ 7: 15–36

Chang ATC, Atwater S, Salomonson VV, Estes JE, Simonett DS (1980) L-band radar sensing of soil moisture. NASA Rep NASA-TM-80628

Churchill PN, Keech MA (1983) SAR investigations of Thetford forest. Europ SAR-580 Invest Prelim Rep (ESA)

Churchill PN, Horne AID, Kessler R (1985a) A review of radar analyses of woodland. Proc EARSeL Workshop Microwave remote sensing applied to vegetation, Amsterdam, 10–12 Dec 1984. ESA SP-227, Jan 1985, pp 25–32

Churchill PN, Deane GC, Trevett JW (1985b) An evaluation of SAR-580 multi-frequency radar data over the Norfolk test site (GB4). The European SAR-580 Experiment; Invest Final Report pp 671–691

Cimino JB, Brandani A, Casey J, Rabassa J, Wall SD (1986) Multiple incidence angle SIR-B experiment over Argentina: mapping of forest units. IEEE Trans Geosci Remote Sensing GE-24; 498–509

Daus SJ, Lauer DT (1971) Testing the usefulness of SLAR imagery for evaluating forest vegetation resources. Univ Calif, Final Rep, For RSL, Berkeley

Davis CF (1973) Forest vegetation mapping with synthetic aperture radar. ERIM Int Rep

Dobson MC, Ulaby FT (1981) Microwave backscatter dependence on surface roughness, soil moisture, and soil texture. Part III. Soil tension. IEEE Trans Geosci Electron GE-19: 51–61

Dobson MC, Ulaby FT (1986) Active microwave soil moisture research. IEEE Trans Geosci Remote Sens, GE-24: 23–36

Dobson MC, Ulaby FT, Hallikainen MT, El-Rayes MA (1985) Microwave dielectric behaviour of wet soil. Part II. Four-component dielectric mixing models. IEEE Trans Geosci Remote Sens, GE-23: 35–46

EASAMS (1973) SLR systems and their potential application to earth resource surveys. Potential application of SLR. Rep ESRO-CR-138, vol 3

ESA (1987) Simulation of SAR data products. Contract 6.619/85/F/FL/(SC)

Evans DL (1986) Multi-polarization radar images for geologic mapping and vegetation discrimination. IEEE Trans Geosci Remote Sensing GE-24: 246–257

Goodenough DG (1980) Quantitative analysis of SAR data for detecting forest infestations of insects. Final rep Airborne SAR proj. Intera Env Consult. Rep ASP-80-1

Hallikainen MT, Ulaby FT, Dobson MC, El-Rayes MA, Wu L (1985) Microwave dielectric behaviour of wet soil. Part I. Empirical models and experimental observations. IEEE Trans Geosci Remote Sens, GE-23: 25–34

Hardy NE, Coiner JC, Lockman WD (1971) Vegetation mapping with SLAR; Yellowstone National Park. In: Lomax JB (ed) Propagation limits remote sensing, AGARD Conf Proc 90

Hipp JE (1974) Soil electromagnetic parameters as a function of frequency, soil density and soil moisture. Proc IEEE 62: 98–103

Hirosawa H, Matsuzaka Y (1987) Measurement of scattering coefficient, #, of trees. Int J Remote Sensing 8: 609–616

Hirosawa H et al. (1987) Measurement of backscatter from conifers in the C and X bands. Int J Remote Sensing 8: 1687–1694
Hoekman DH (1985) Radar backscattering of forest stands. Int J Remote Sensing 6: 325–343
Hoekman DH (1987) Measurement of the backscatter properties of forest stands at X-, C- and L-band. Remote Sensing Environ 23: 397–416
Hoekstra P, Delaney A (1974) Dielectric properties of soils at UHF and microwave frequencies. Geophys Res 79: 1699–1708
Holmes MG (1987) Monitoring arable crops with radar. Crops 4: 22–23
Holmes MG (1990) Applications of radar in agriculture. In: Steven MD, Clark JA (eds) Applications of remote sensing in agriculture. Butterworths, London, pp 307–330
Holmes MG (1991) Monitoring vegetation in the future: radar. Bot J 108: 93–109
Hoogeboom P (1982) Classification of agricultural crops in radar images. IGARSS, vol 2, FA-4. Munich
Hoogeboom P (1983) Classification of agricultural crops in radar images. IEEE Trans Geosci Remote Sensing GE-21: 329–336
Horne AID, Rothnie B (1985) The use of optical SAR-580 data for forest and non-woodland tree survey. The European SAR-580 Experiment; Investigators Final Report, pp 509–523
Hunting Geology and Geophysics (1981) Final Report of the data content of overland SEASAT-A SAR imagery. Report to client
Kessler R, Reichert P, Losche P (1983) Feasibility of different evaluation method with X and C-band SAR-CV-580 data in agricultural and forestry test sites. SAR-580 Invest Prelim Rep (ESA)
Knowlton DJ, Hoffer RM (1981) Radar imagery for forest cover mapping. Proc 7th Symp on Machine processing of remotely sensed data with special emphasis on range, forest and wetlands assessment. IEEE Computer Soc, pp 626–632
Koolen AJ, Koenigs FFR, Bouten W (1979) Remote sensing of surface roughness and top soil moisture of bare tilled soil with an X-band radar. Neth J Agric Sci 27: 284–296
Lamont J et al. (1987) Ground truth requirements for radar observations over land and sea. Int J Remote Sensing 8: 1057–1067
Lee J (1980) SAR for identifying, monitoring and predicting conditions in the forest environment in British Columbia. Final Rep Airborne SAR Proj, Intera Env Consult Rep ASP-80-1
Le Toan T (1982) Active microwave signatures of soil and crops. Significant results of three years of experiments. IEEE Publ 82 C414723 TP-2: 3.1–3.5
Le Toan T, Shahin A, Riom J (1980) Application of digitalized radar images to pine forest inventory: first results. Proc. 14th Int Symp on Remote Sensing Environment, San Jose, Costa Rica, vol 2, pp 943–944
Le Toan T, Flouzat G, Pausander M, Fluhr A (1981) Soil backscatter experiments in the 1.5 to 9 GHz region. Proc URSI Comm F of Signature Prob Microwave Remote Sensing Earth, Lawrence, KS
Le Toan T, Lopez A, Huet M (1984) On the relationships between radar backscattering coefficient and vegetation canopy characteristics. Proc. IGARSS 84 Symp, Strasbourg, pp 155–160 ESA SP-215
Lewis AJ (1971) Cumulative frequency curves of the Darien Province, Panama. In: Lomax JB (ed) Propagation limits in remote sensing. AGARD Conf Proc 90
Li FK, Ulaby FT, Easton JR (1980) Crop classification with a Landsat-radar combination. Proc Symp on Mach process remote sens data. Purdue Univ West Lafayette
de Loor GP (1983) The dielectric properties of wet materials. IEEE Trans Geosci Remote Sensing GE-21: 364–369
de Loor GP, Hoogeboom P, Attema EPW (1982) The Dutch ROVE program. IEEE Trans Geosci Remote Sensing GE-20: 3–11
MacDonald HC, Waite WP, Damariche JC (1980) Use of Seasat satellite radar imagery for the detection of standing water beneath forest vegetation. Am Soc Photogramm Tech Meeting Niagara Falls
McDonald KC, Dobson MC, Ulaby FT (1990) Using MIMICS to model L-band multi-angle and multi-temporal backscatter from a walnut orchard. IEEE Trans Geosci Remote Sensing GE-28: 101–109
Megier J, Mehl W, Ruppelt R (1985) Methodological studies of land use classification of SAR and mulit-sensor imagery. The European SAR-580 Experiment; Investigators Final Report pp 693–722

Morain SA (1967) Field studies on vegetation at Horsefly Mountain, Oregon, and its relation to radar imagery. Univ Kans Cent Res. Lawrence, Kansas, RSL Tech Rep, pp 61–62
NASA Earth Observ System (1987) Instr Panel Rep, vol IIf. SAR
Newton RW (1977) Microwave remote sensing and its application to soil moisture detection. Remote Sensing Cent, Texas A&M Univ, Coll Stn Tech Rep RSC 81
Newton RW, Rouse JW (1980) Microwave radiometer measurements of soil moisture content. IEEE Trans Antennas Propag AP-28: 680–686
Njokyu EG, O'Neill PE (1982) Multifrequency microwave radiometer measurements of soil moisture. IEEE Trans Geosci Remote Sensing GE-20: 468–475
Pala S (1980) Radar imagery for assessment of coniferous forest regeneration. Final Rep Airbor SAR proj, Intera Env Consult, Rep ASP-80-1
Paris JF (1982) Radar remote sensing of crops. IGARSS vol 2, Munich
Paris JF (1986) The effect of leaf size on the microwave backscattering of corn. Remote Sensing Environ
Parry JT (1974) X-band radar in terrain analysis under summer and winter conditions. Proc 2nd Can Symp Remote sensing, vol. 2. pp 4721–4786, Guelph, Ontario
Parry JT (1977) Interpretation techniques for X-band SLAR. Proc 4th Can Symp on Remote sensing, Quebec
Peterson RM, Cochrane GR, Morain SA, Simonett CC (1969) A multisensor study of plant communities at Horsefly Mountain, Oregon. In: Johnson P (ed) Remote sensing ecology, Georgia Univ Press, pp 63–93
Richards JA et al. (1987) L-band radar backscatter modeling of forest stands. IEEE Trans Geosci Remote Sensing GE-25
Richards JA (1987) An explanation of enhanced radar backscatter from flooded forests. Int J Remote Sensing 8, pp 1093–1100
Rubec C (1980) Ecological land classification using radar imagery. Final Rep Airborne SAR Proj, Intera Env Consult, Rep ASP-80-1
Schmugge TJ (1978) Remote sensing of surface moisture. J Appl Meteorol 17: 1550–1557
Schmugge TJ (1980) Effect of texture on microwave emission from soils. IEEE Trans Geosci Remote Sensing GE-18: 353–361
Schmugge TJ (1983) Remote sensing of soil moisture: recent advances. IEEE Trans Geosci Remote Sensing GE-21: 336–344
Schmugge TJ, Jackson TJ, McKim HL (1980) Survey of methods for soil moisture determination. Water Resour Res 16: 961–979
Shanmugan KS, Ulaby FT, Narayanan V, Dobson MC (1983) Identification of corn fields using multi-date radar data. Remote Sensing Environ 13(3)
Shuchman RA, Lowry RT (1977) Vegetation classification with digital X-band and L-band dual polarized SAR imagery. Proc 4th Can Symp Remote sensing. Quebec, pp 444–458
Sicco-Smit G (1974) Practical application of radar images to tropical rainforest mapping in Colombia. Mitteilungen der Bundesforschungsanstalt fuer Forst- und Holzwissenschaft 99: 51–64
Sieber AJ (1985) DFVLR participation in the European SAR campaign. The European SAR-580 Experiment; Investigators Final Report, pp 735–738
Sieber AJ, Freitag B, Lawler K (1982) The aspect angle dependence of SAR images. IGARSS vol 2, FA-4, Munich
Simonett DS (1978) Potential applications of space radar for agricultural monitoring in the wet tropics. Proc 11th Int Symp on Remote Sensing of the environ. Ann Arbor, vol 2, pp 427–428
Taube D, Theiss SW (1984) Correlation of microwave sensor returns with soil moisture. Proc IGARSS '84 Strasbourg, pp 251–256
Ulaby FT (1980) Microwave responses of vegetation. Adv Space Res 1: 55–70
Ulaby FT, Bare JE (1979) Look direction modulation function of the radar backscattering coefficient of agricultural fields. Photogramm Eng Remote Sensing 45: 1495–1506
Ulaby FT, Batlivala PP (1976a) Optimum parameters for mapping soil moisture. IEEE Trans Geosci Electron GE-14: 81–93
Ulaby FT, Batlivala PP (1976b) Diurnal variations of radar backscatter from a vegetation canopy. IEEE Trans Antennas Propag AP-24: 11–17

Ulaby FT, Bush TF (1976) Monitoring wheat growth with radar. Photogramm Eng and Remote Sensing 42: 557–568

Ulaby FT, Batlivala PP, Dobson MC (1978) Microwave backscatter dependence on surface roughness, soil moisture, and soil texture. Part I. Bare soil. IEEE Trans Geosci Electron, GE-16: 286–295

Ulaby FT, Bradley GA, Dobson MC (1979) Microwave backscatter dependence on surface roughness, soil moisture, and soil texture. Part II. Vegetation-covered soil. IEEE Trans Geosci Electron GE-17: 33–40

Ulaby FT, Li RY, Schanmugan KS (1981) Crop classification by radar. Univ Kans Cen Res, Lawrence, Kans, RSL Tech Rep 360-13

Ulaby FT, Aslam A, Dobson MC (1982) Effects of vegetation cover on the radar sensitivity to soil moisture. IEEE Trans Geosci Remote Sensing GE-20: 476–481

Ulaby FT, Moore RK, Fung AK (1986) Microwave remote sensing active and passive. Artech House, Norwood

Ulaby FT et al. (1986) Textural information in SAR images. IEEE Trans Geosci Remote Sensing GE-24: 235–245

Wang JR, McMurtrey JE, Engman ET, Jackson TJ, Schmugge TJ, Gould WJ, Fuchs LE, Glazar WS (1982) Radiometric measurements over bare and vegetated fields at 1.4 GHz and 5 GHz frequencies. Remote Sensing Environ 12: 295–311

Wang JR, Engman ET, Shiue JC, Ruzek M, Steinmeier C (1986) The SIR-B observations of microwave backscatter dependence on soil moisture, surface roughness and vegetation cover. IEEE Trans Geosci Remote Sensing GE-24: 510–515

Westman WE, Paris JF (1987) Detecting forest structure and biomass with C-band radar: physical model and field tests. Remote Sensing Environ 22: 249–269

Wobschall D (1977) A theory of the complex dielectric permittivity of soil containing water; the semidisperse model. IEEE Trans Geophys Electron, GE-15: 49–58

Wu ST (1983) Analysis of synthetic aperture radar acquired over a variety of land cover. IGARSS 2, FP-5, San Francisco

Wu ST (1986) Preliminary report on measurements of forest canopies with C-band radar scatterometer at NASA/NSTL. IEEE Trans Geosci Remote Sensing GE-24: 894–899

Wu ST (1987) Potential application of multipolarization SAR for pine-plantation biomass estimation. IEEE Trans Geosci Remote Sensing GE-25: 403–409

Wu ST, Sader (1987) Multipolarization SAR data for surface features delineation and forest vegetation characterization. IEEE Trans Geosci Remote Sensing GE-25: 67–76

Zoughi et al. (1986) Identification of major backscattering sources in trees and shrubs at 10 GHz (X-band). Remote Sensing Environ 19: 269–290

Subject Index

Springer-Verlag and the Environment

We at Springer-Verlag firmly believe that an international science publisher has a special obligation to the environment, and our corporate policies consistently reflect this conviction.

We also expect our business partners – paper mills, printers, packaging manufacturers, etc. – to commit themselves to using environmentally friendly materials and production processes.

The paper in this book is made from low- or no-chlorine pulp and is acid free, in conformance with international standards for paper permanency.

GPSR Compliance
The European Union's (EU) General Product Safety Regulation (GPSR) is a set of rules that requires consumer products to be safe and our obligations to ensure this.

If you have any concerns about our products, you can contact us on

ProductSafety@springernature.com

In case Publisher is established outside the EU, the EU authorized representative is:

Springer Nature Customer Service Center GmbH
Europaplatz 3
69115 Heidelberg, Germany

www.ingramcontent.com/pod-product-compliance
Ingram Content Group UK Ltd.
Pitfield, Milton Keynes, MK11 3LW, UK
UKHW021951270726
14060UKWH00002B/469

* 9 7 8 3 5 4 0 5 6 3 8 1 5 *